Chronological Developments of Wireless Radio Systems before World War II

Vinayak Laxman Patil

Chronological Developments of Wireless Radio Systems before World War II

 Springer

Vinayak Laxman Patil
Research Centre of Maharashtra
Institute of Technology (MIT)
Dr. Babasaheb Ambedkar
Marathwada University
Aurangabad, Maharashtra, India

ISBN 978-981-33-4904-9   ISBN 978-981-33-4905-6   (eBook)
https://doi.org/10.1007/978-981-33-4905-6

This Springer imprint is published by the registered company Springer Nature Singapore Pte Ltd.
The registered company address is: 152 Beach Road, #21-01/04 Gateway East, Singapore 189721, Singapore

To my wife, Vidya

Preface

From the ages, human race has strived hard for better and better communication with their fellow colleagues by adopting various novel ways leading to the current state of wireless radio communication systems. It is remarkable to present the motivating, exciting consolidated account of chronologies of discoveries, inventions and developments put forth by various stakeholders in the area of wireless radio communication systems. It was the curiosity of inventors that made it happen, which can be judged from the following quotes:

> If you want to find the secrets of universe, think in terms of energy, frequency and vibrations.....—Nicola tesla

And this is what exactly happened while investigating the phenomenon called "lightning and thunder" that amounts to the investigation of energy, frequency and vibration. A spark was responsible for making inventors investigate the type of energy released from it, called electromagnetic waves and became the natural choice for inventors like Maxwell and Hertz. James Clerk Maxwell speculated the existence of electromagnetic waves and how power transfer through free space occurs and put forward a theory called "Electromagnetic Field Theory". The quotation of James Clerk Maxwell has something to tell about the fascination of discoveries of science, and focuses on the chronological tracing of the wireless radio systems to make them fascinating.

> In Science, it is when we take some interest in the great discoverers and their lives that it becomes endurable, and only when we begin to trace the development of ideas that it becomes fascinating.—James Clerk Maxwell

Now it was the turn of Heinrich Rudolf Hertz to practically prove the existence of Electromagnetic Waves: he connected a large area plate to spark gap to increase charge holding capacity so that spark can be produced across the spark gap by adjusting the gap, and the radiation emitted was sensed using loop antenna and was no sure about its use in future.

I do not think that radio waves I have discovered will have any practical application...
—Heinrich Hertz

This quote indicates that Heinrich Rudolf Hertz was unaware of the implications of his discovery and could not even foresee the importance and impact of his discovery as seen today. It now very well proved that wireless systems are extensively using electromagnetic waves.

When wireless is perfectly applied, whole earth can be converted into a huge brain and we shall be able to communicate with each other instantly irrespective of distance.—Nicola Tesla

Guglielmo Marconi worked tirelessly for large scale public spread of wireless radio communication systems and commented on his involvement in the wireless radio communication systems and fuelled never-ending quest or race in the wireless radio communication system developments.

Have I done the world good, or have I added a menace?—Guglielmo Marconi

Reginald Aubrey Fessenden introduced brilliant ideas for sound transmission and entirely focused on telephony and kept the ball rolling further. Many others then jumped into the fray for developing radio transmitter, receivers and required components.

... in going over the history of all the inventions for which history could be obtained it became more and more clear that in addition to training and in addition to extensive knowledge, a natural quality of mind was also necessary.—Reginald Fessenden

This book not only gives running commentary of historical happenings in the area of wireless radio communications, but also gives the detailed understandable technical account of the systems in great detail, not only for students, but also for the readers of all segments. Efforts are taken so that the information provided in the book is from original resources of inventors, and in case of absence of authenticated information from original inventors, the information provided is verified using multiple references. And on the lighter side, finally, I will end up this preface with my own signature quote,

Quotes of only celebrities are admired and referred to for value addition, while even best quotes of common man go unnoticed...—by Author

The Author

Acknowledgement

I sincerely thank my critical son Dr. Nilesh Patil for providing me a series of required humiliating triggers for my slow but iterating perfectionist approach: he provided the required acceleration and speed for manuscript completion. I greatly oweto his valuable contribution into the preparation of this book.

I greatly appreciate the encouragement given by my elder son Dr. Vinit Patil and his wife Stacy Patil as well as their son Veer and daughter Mala during the course of preparation of this manuscript.

I also thank Dr. Santosh Bhosale, Principal, and Dr. Arun Sabale, Head Research Centre, Maharashtra Institute of Technology, an affiliate of Dr. Babasaheb Ambedkar Marathwada University, Aurangabad (India), for encouraging me to write such a book for the benefit of the readers of all segments.

Thanks are also due to Dr. Madhukar Deshmukh, Dr. Arun Sabale and Dr. Vaibhav Hendre for critically going through the manuscript of this book.

My parents, Late Shri Laxman Patil and Mrs. Manakarnika Patil, besides their poor financial farming background, kept my temperament all time high during my whole education and job carrier, helped me to a maximum extent and did all those things that they can do for me.

Last but not least, I also thank my wife Vidya for her patience and tolerating me for my non contribution to home and family affairs during the course of preparation of this manuscript and I dedicate this book to her.

The Author

Contents

About the Author

Dr. Vinayak Laxman Patil is affiliated with the Research Centre of Dr. Babasaheb Ambedkar Marathwada University, Maharashtra, India. He has previously served at several research and academic institutions such as Central Electronics Engineering Research Institute, Pilani, India, and Director of institutes like Centre for Electronics Design & Technology (CEDT) and Mahatma Gandhi Missions's Infotech in Aurangabad, India. He has done his M.E. in Electronics and Telecommunications and Ph.D. from the University of Pune. He has received numerous awards over the years and is a member of many professional societies, including IETE, IEEE, Instrument Society of America, etc.

List of Figures

List of Tables

Chapter 1
Ancient Wireless Communications

Abstract The present chapter describes the ancient forms of communications adopted by animals and humans to send distress signals to their associates making typical sounds or some short symbols for immediate help which was then evolved for longer distances. Long distance communication using symbols was evolved as "Telegraphy" and long distance communication using phonetic sounds was then evolved as "Telephony". The combination of both was further evolved as "Television" or "Tele-Videophony". Due to human eyesight limitations, it was a challenge about "How to send this information to longer distances". With the inventions of various forms of electricity, these problems were also comfortably solved. The representative information is usually represented by the symbols or textual form or voice. Such information was converted into electric signals for transmission to the longer distances using either wired or wireless media, while at the receiving end, electric signals were decoded to get the original message back.

1.1 Introduction to Long Distance Communication

From the very olden days, animals and humans use to send distress signals to their distant associates for getting their help in the need of hour. The signals such as flags, typical phonetic voices, or actions were used, and hence the field of telecommunication can be divided into two parts, one part deals with the transmission of conceivable *information* may be textual messages, letters or alphabets, coded messages that can be interpreted at receiving end called *telegraphy*, while another type is long distance transmission of the voice itself, which is called as *telephony*. In today's scenario both symbolic information and voice signal are converted to their electrical equivalent that can be suitably transmitted over wired or wireless media to longer distances. This could happen because of intelligence possessed by the humans beings, they started exploring novel ways to send voice and information signals to longer distances, and as a result, you can see the great evolution of sophisticated systems like the internet applied for both the type of applications.

V. Patil, *Chronological Developments of Wireless Radio Systems before World War II*,
https://doi.org/10.1007/978-981-33-4905-6_1

It all started with symbolic language for conveying specific information that was of interest and concern to their daily livelihood. Each discretely symbol conveyed specific information and since telegraphy deals with information it was ahead of developments of Telephony. The signal information may consist of visual information, audio information or audio-visual information and even data information. The branch that deals with sending voice signals to longer distances is called Telephony, while the branch that deals with sending both audio and visual information together is called as video signal to the longer distances is called as Tele-Videophony and yet another branch that deals with sending numerical information or data signals to longer distances is Telegraphy, while "Telecommunication" is a general designation for all these types of long distance communications.

1.1.1 Wireless Telegraphic Communication Methods

Since telegraphy was set to represent human signals using discrete symbols and since discrete signals were easy for communication as compared to continuous representing voice signals and hence the spread of wireless Telegraphy communication was rapid. This chapter specifically deals with various aspects of wireless telegraphic systems. The word telegraphy is derived from Greek words like tête means "distance" and grä̈phein means "to write". Telegraphy paved a way for the evolution of other types of communications.

The basic principles used in conveying the information or messages, used devices based on visual patterns of information (optical), electrostatic and relays operating on electromagnetic principles as shown in Fig. 1.1.

The Fig. 1.1 a represents the symbols those can be visible from suffciently large distance and interpreted by a person at the far end. In Fig. 1.1b, when the capacitor is

Fig. 1.1 The principles adopted for sensing of symbols in various types of telegraph instruments

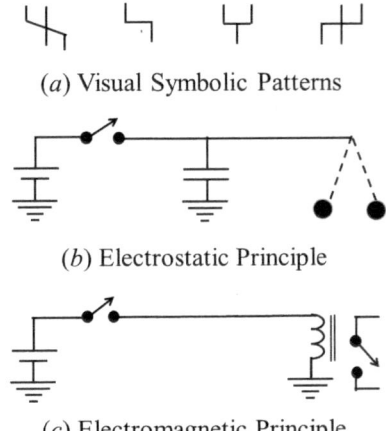

(*a*) Visual Symbolic Patterns

(*b*) Electrostatic Principle

(*c*) Electromagnetic Principle

charged, two hanging balls also get charged and repel each other and move apart and if the capacitor is discharged, then the balls move closer. In Fig. 1.1c, if the switch is closed, current flows through the coil and contact is magnetically closed, while if the switch is opened, coil gets de-energised and contact is opened.

1.2 Origin and Very Early Forms of Communications

Smoke and light signals have been the oldest forms of visual signalling [1] for sending the distress signal for getting help from the communities staying at longer distances and dates back _588 BC_ as reported in the earliest reliable record of writings of Jeremiah by communicating intelligence to others by the "signs of fire". Homer was the first person to make a mention of telegraphic art, and the Roman generals took the help of such methods for distant communications during wartime.

During _BC 264_, Polybius wrote "Punic Wars" where in he mentions that he was able to improve communication by use of "Communicating ideas by letters" which was used by Cleoxenus in his telegraph. During _BC 1084_, Troy city was besieged by Agamemnon, who signalled this event to his queen Clytæmnestra.

1.2.1 Aerial Telegraphy Using Semaphores

The visual telegraph system was used in Europe and Asia for conveying the letters in symbolic fashion or in the form of patterns that were communicated visually. These systems were then superseded by electric telegraph systems.

1.2.2 Claude Chappé's Optical Telegraph

In _1794_, Claude Chappé [2–4], French origin inventor, worked hard to successfully establish first aerial telegraph line using line of sight communication between Paris and Lille. It was, in fact, optical telegraph and semaphore or flag based alphabets that released information from one hilltop to other. It used two movable arms to represent various position of arms to represent visual information that codified various alphabets as shown in Fig. 1.2. However, such systems could not be used in foggy environment.

1.2.2.1 Limitations of Optical Telegraph Communications

The optical telegraphic communications were not free from drawbacks due to the limited ranges of human eyesight posing serious limitations to the communication.

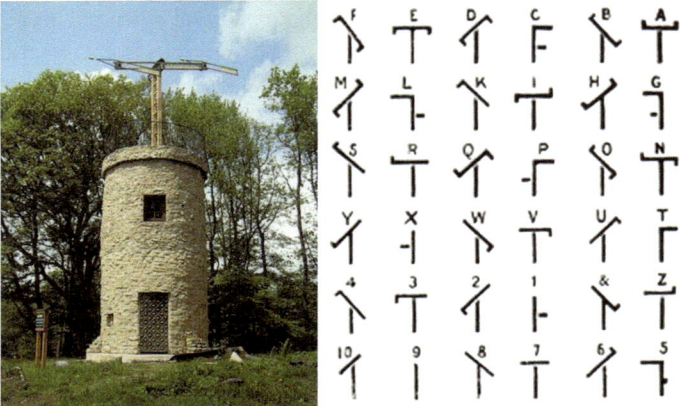

Fig. 1.2 Claude Chappé semaphore or flag based optical telegraph. *Credit* CC-BY-SA 3.0 Lokilech in der Wikipedia auf Deutsch | 2006 | Optischer Telegraf (Replikation) von Claude Chappe auf dem "Litermont" bei Nalbach in Deutschland

At this point of time, people were aware of some phenomenon of static electricity occurring in nature and realised that static electricity can be used for producing the spark and pattern of sparks thus produced can be used to identify the information or symbols.

1.3   Discoveries of Various Forms of Electricity and Magnetism

The static electricity played a crucial role in the development of telegraphy in the early days. The origins of static electricity [5] can be traced back to the work of Thales of Miletus in *BC 585*, who observed that loadstone attracts iron with a belief that loadstone has soul or God in it. However, Democritus during *(circa 460BC to circa 370BC)*, an influential pre-Socratic Greek philosopher known for atomic theory of the universe in which atoms followed natural laws. The Epicurus *(circa 342BC to circa 270BC)* and Democritus believed that the soul consists of atoms that disintegrate after death. In *circa 300BC*, Theophrastus observed that some substances have static electricity when rubbed. However, both Thales of Miletus and Theophrastus could not explain the phenomenon, but they realised that some substances attract each other while some substance repel each other.

William Gilbert studied at St. John's College, Cambridge and became a doctor and personal physician of Queen Elizabeth. He published a book called "De Magneto" in *1600*, opened the doors of science in magnetism and could predict that earth is a giant magnet. He started the investigation of electrical phenomenon using electroscope and noticed that both electrical and magnetic attractions become stronger when objects get closer and closer.

Charles Francois de Cisternay DuFay, a French scientist, believed that there are two types of electricity called "Vitreous (Produced by glass, precious stones, crystals, hair, etc.)" and "Resinous (Produced by resins)" linked to the substances those produce it. His observations were published in *1733* [6, 7].

Leyden jar was the first device that was invented by Ewald G. von Kleist, on *November 4, 1745* that could store electric charge. Ewald G. von Kleist nearly his invention that was credited to Pieter van Musschenbroek of Leyden, Holland in *1746*.

Benjamin Franklin discovered the identity of lightning and electricity, who experimented in *June 1752* using kite and key attached to the lower end of the string, he observed a bright spark between key and finger inventing static electricity.

1.4 Proposed Ideas, Thoughts and Suggestions for Telegraphic Communication Devices

The first and foremost suggestion of electric telegraph was came from an anonymous person called "C.M." (later on he was most likely to be identified as Charles Morrison, of Greenock, who migrated to Virginia where he died) [3, 8, 9] was living in a town in Scotland called Renfrew, who published his work in Scots Magazine by his own identity as "C.M." [10], in *February 17, 1753*. It consisted of insulated wires running from one end to another, wherein one wire was reserved for one alphabet and wires were excited one at a time using electricity depending on alphabet to sent.

1.4.1 Bozolus's Telegraph

In *1767*, Joseph Bozolus, lecturer of philosophy at a college in Rome, was next to suggest telegraph based on sparking principle [11] in which he proposed laying of wires between two stations without touching each other. Letters in the form of sparks were identified.

1.4.2 Maunoir Odier's Idea About Telegraph

In *1773*, Maunoir Odier, physicist of Geneva [11], in his communication to a lady of his acquaintance said, I have an idea about communication anywhere in any language within half an hour.

Fig. 1.3 Georges Louis
Lesage electric telegraph,
Source Louis Figuier, Les
Merveilles De La Science 2,
867 - 1891, Tome 1.djvu, Le
premier télégraphé électrique
(apsareil de Georges Lesage,
executé à Geneve en 1771)

1.4.3   George Louis Le Sage's Electric Telegraph

In *1774*, George Louis Lesage [2, 3, 12], a philosopher of French origin realised early electric telegraph (Fig. 1.3) in Geneva to cater the communication within two rooms of his house that consists of one insulated wire for each of 26 letters of alphabets. When wire is electrically energised, the ball suspended at other end was moved to strike a bell to produce to identify which alphabet was transmitted.

1.4.4   Volta's Electric Telegraph

In *15th April, 1777*, Alessandro Volta reported his telegraph [13] along with experimental details to Professor Barletti, although this proposition was forestalled by Charles Marshall, of Renfrew in *1753*.

1.4.5   Invention of Battery

In *1780*, an Italian anatomist Luigi Galvani, constructed a crude battery using two dissimilar metals and natural fluid from dissected frog.

1.4.6   Anonymous Telegraph

In *1782*, it was an anonymous letter posted to "Journal de Paris" [14] that described glowing letters due to discharge between metallic foils.

1.4.7   Lomond's Electric Telegraph

Monsieur Lomond in *October 16, 1787*, invented a telegraph [15] to communicate between electrical machine and electrometer located in neighbouring rooms, respectively, the electrometer was constructed using pit balls. The equipments were connected using two wires, at each discharge of the Leyden, balls recede or move away till they come in contact with return wire. Single divergence was treated as letter A, two divergences were treated as letter B and so on. The working principle of Monsieur Lomond's telegraph is depicted in Fig. 1.1b.

1.4.8   Réveroni–Saint-Cyr's Telegraph

In *1790*, Réveroni–Saint-Cyr proposed electric telegraph [11] and declared the results of lottery using his telegraph.

1.4.9   Reusser's Telegraph

In *1794*, Reusser's invented electric telegraph [11] that was clumsy as compared to the telegraphs of both Lomond and Chappé, Reusser claimed that he was able to dictate a complete letter from his house to the person sitting at other end using luminous planes connected to the sending side by wires.

1.4.10   Böckmann, Lullin and Cavallo's Telegraphs

In *1794*, all of these inventors proposed modifications to the Reusser's plans [11, 16–18], these methods involved modifications made to his initial codes using sparks and intervals.

1.4.11   Reizen's Electric Spark Telegraph

Reizen used electric spark for his telegraph equipment as reported by Voigt's Magazine [19] in *1794*, He arranged 26 letters related to alphabets and 10 numeral to make 36 in total. These letters are arranged in the form of 6 × 6 matrix having 6 rows and 6 columns. Each row having 6 patterns. Each pattern is separated or spaced by an area of 1 in. square. The patterns formed of tin foil strips having two ends with the one end marked as P, which means positive (+) while the other end is marked as negative N

Fig. 1.4 Reizen's electric spark telegraph

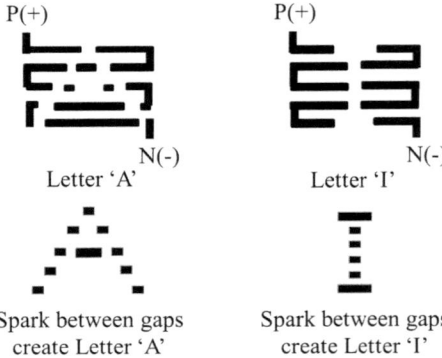

(−). It used 36 wires for 36 alphabets and one wire for common point. Each pattern is not continuous but interrupted depending on shape and visual appearance of letter as shown in Fig. 1.4.

1.4.12 Cavallo's Signal Coding in the Form of Number of Sparks for Telegraphs

Cavallo in *1795*, used number of sparks to designate signals and explosion of gas is used as alarum (alarm).

1.4.13 Salvá's Electric Spark Telegraph

In *1795*, similar to Reizen's telegraph, Don Francisco Salvá, of Barcelona [11, 20], constructed his own telegraph, His proposal to Academy of Sciences, Barcelona on *December 16, 1795* mentions regarding two wires one from Barcelona to Mataro and others from Mataro to Barcelona, when connected to Leyden Jar, the persons, holding the wires at both the ends got shock.

The report also says "If electricity is to be of any use, it must be able to communicate the information". It also said every word can be expressed with sufficient accuracy using 22 letters or even 18 letters. A man could communicate with a other person in Mataro using 44 wires and could identify the combination of wires on which person gets shock could identify the letters.

Salvá was the first person to propose a galvanic telegraph. From *1796* to *1799*, Salvá was invited by the Academy of Science of that capital for the experiments of public interest, and hence resided in Madrid during that period. According to Salvá's friend, Dr. Balcells, the telegraph requiring single wire was constructed in *1798* that was used between Madrid and Aranjuez covering a distance of twenty six miles.

In *1834*, the remains of Salvá's telegraph were presented to the College of Pharmacy of San Fernando, where Dr. Balcells was Adjutant, the remains of Salvá's telegraph were earlier destined to be in Don Antonio's museum.

1.5   Conclusion

This chapter put forward the ancient means of communications and some of the ideas put forward by various philosophers during that period for long distance communications including the use of various forms of electricity produced by natural methods.

Chapter 2
Search for Wireless Power Transfer Methods

Abstract This chapter covers various power transfer mechanisms for both wired and wireless communications. Although the focus of this book relates to the chronology of developments of wireless systems, some of the concepts of wired systems are also applicable to the wireless systems as the basic principles of operations are same for both type of these systems. The battery is the fundamental source of power for all types of communication circuits and was an important invention that provided continuous electric power to communication circuits. In long distance communication, it was a challenge for transferring the electric power between transmitting and receiving ends. The basic principles and methods of electric power transfer are described in this chapter. The electric spark investigated by Benjamin Franklin had an important and critical place in communication systems that led to the invention of electromagnetic waves now widely used in wireless communication.

2.1 Introduction to Wireless Power Transfer

The discoveries of various forms of electricity and magnetism were the turning points in the history of wireless communications. It was the lightning in the sky that fascinated many intuitive philosophers and scientists.

2.1.1 Benjamin Franklin's Experiment

The roots of today's sophisticated wireless developments trace down to the quest of human beings to understand the various phenomenons occurring in nature. One of such phenomenons called thundershower or thunderstorm or lightning is characterised by change of one form of energy into another form of energy in which electrical energy is converted into light and sound energy. These forms of energies do exist in nature and nobody can be credited as an inventor of such energies but somebody who keenly observed and studied such phenomenon should be credited with its discovery.

In *June 15, 1752*, Benjamin Franklin [6, 21] conducted dangerous kite and key experiment in rainy and cloudy weather to know what exactly is the phenomenon of lightning and thundering. While conducting his experiment, on touching the key he got shock with bright spark between his finger and key. His experiment proved that lightning and sparks are the same. Earlier scientists experimented only static electricity, Benjamin Franklin's work was one step ahead to predict that electricity has positive and negative elements and electricity flows between them.

2.1.2 Use of Sparks Produced by Static Electricity in Communication

The early spark communication ideas described in the previous chapter were based on the use of static electricity that required electrostatic energy storage mechanisms prone to quick power discharge and hence it was an ideal time for the invention of electric power sources that can supply continuous power to communication devices. The symbols were based on electrostatic charges and sensing resulting potentials at far end of communication. The current sensing communication symbols were hardly employed due to limitations of availability of power sources that can produce continuous current in an electric circuit and hence instead of using charge storage devices prone to quick discharge, voltage sources like batteries were badly needed.

2.1.3 Invention of Battery

The Scientists then directed their efforts in the direction of how to construct the source of electrical energy and in *1800*, Alessandro Volta constructed a device than can produce electric current [22] and then many people performed a series of experiments using this device capable of producing electric current. The device was made of a series of discs of silver and zinc linked in cards soaked in salt water as shown in Fig. 2.1.

Volta's invention provided a reliable and long life continuous source of electrical energy as compared to electrostatic storage mechanisms those can only be used for short durations and prone to loose the electric energy even when energy is not used and because of these properties, electrostatic storage mechanisms were never prefered as sources of electric energy.

(a) (b)

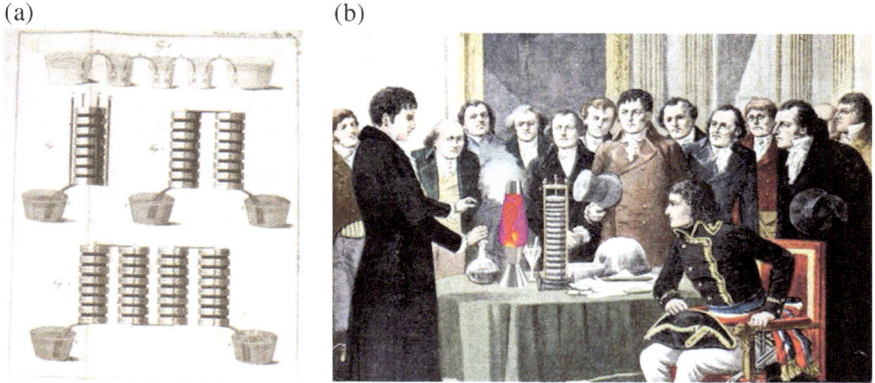

Fig. 2.1 Volta's battery **a** Plate from Volta's landmark book: on the electricity excited by the mere contact of conducting substances of different kinds 1800, Credit: SPARKMUSEUM **b** Demonstration Volta's battery to Napoleon Bonoparte in 1801, Credit: CC BY 2.0 Mike Licht |2015| NotionsCapital.com Volta demonstrates the electric battery to Napoleon

2.1.4 Basic Principles of Wireless Power Transfer

To remove the jugglery of wires, it was an era for the search of methods that can provide wireless transmission of electric signal generated by some physical process like human voice and how this electric signal containing human voice information can be transmitted without wires.

The wireless power transfer can be primarily classified as electric field coupling, magnetic field coupling and electromagnetic radiation mode. The electric and magnetic fields are mostly coupled through capacitances and inductances, respectively, and are used in short distance wireless energy transfer. The power transfer due to electric field coupling is shown in Fig. 2.2.

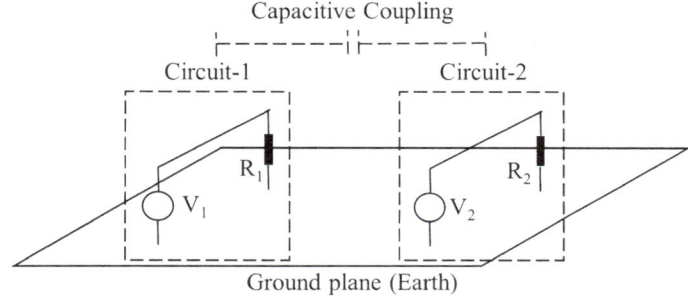

Fig. 2.2 Power transfer through electric field transfer

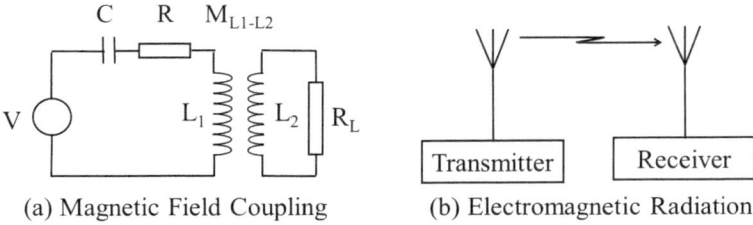

(a) Magnetic Field Coupling   (b) Electromagnetic Radiation

Fig. 2.3 Power transfer through **a** magnetic field and **b** electromagnetic radiation transfer

The power transfer due to magnetic field coupling [23] and by electromagnetic wave radiation [24] are shown in Fig. 2.3a, b.

2.2 Earth Return Path for Reduction of Wires in Long Distance Communications

The early electric telegraphic communications used metallic wires for transmission of symbols over long distances and used batteries to provide power to the telegraphic circuits. The battery's potential was capable of driving the currents through the telegraphic circuits. Since all distributed telegraphic circuits are connected between positive and negative terminals of the battery, all circuit must return to the negative point of the battery, and hence the return conductor required a large cross-sectional area to control large in-rush current at the negative terminal of the battery, and hence people started providing large area for the return path conductor usually referred to as ground plane.

The long distance transmission of telegraphic symbols also required two wires leading to jugglery of wires and the problem of large number of wires to some extent was solved by Carl August Steinheil who put forward an idea of use of earth as return path [25, 26] as a part of circuit to be used in wired communication in *1838*. The Steinheil's concept was later applied to wireless communication circuits. The Steinheil's ground return path is shown in Fig. 2.4.

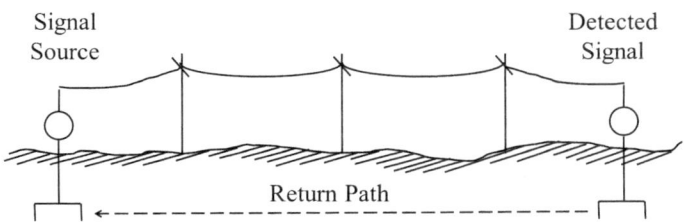

Fig. 2.4 Steinheil's idea of return path for wired circuits

Fig. 2.5 Morse's
experiment of wireless
conduction across river

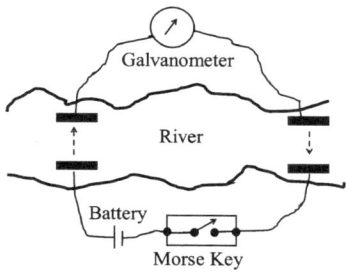

2.2.1   Samuel Morse's Experiment of Wireless Power Transfer

Samuel Morse in *1842*, exploited a principle of wireless conduction through conducting medium and demonstrated that current can be passed without wire through
the river to operate Galvanometric measuring devices at the other end of the river as
shown in Fig. 2.5.

2.3   Electric Spark's Connection with Electromagnetic Waves and Utility in Wireless Communication

Today, we are very well aware of the role of electromagnetic waves and their use
in wireless communication. The high frequency electromagnetic waves are widely
used as carrier waves to carry baseband signal that signifies voice and data signal
that can be compared with telephony and telegraphy, respectively.

2.3.1   Generation of Spark Due to Static Electricity Charges

After the invention of Volta's device, it was a time to conveniently use Volta's device
to generate electric spark and study its properties. The definition of electric spark
comes from the static electricity experiments of rubbing glass tube with dry silk in
dark that produces a cracking sound and sparks of intense light [27], these sparks of
intense light emitted in dark is called as "Electric Spark", which provided a spark of
inspiration to the researchers.

2.3.2   James Clerk Maxwell's Speculations About Electromagnetic Waves

In 1864, James Clerk Maxwell was the first person to speculate that light from spark
is a traverse electromagnetic wave [28] and further existence of electromagnetic

waves was practically demonstrated by Heinrich Rudolf Hertz in _1897_. His experiment proved that electric spark generates the waves that he was able to detect and were named after him "Hertzian Waves". His experiment also proved that electromagnetic waves can be sensed at longer distance and hence can be used in wireless communication, which is described in detail in following section (sect. 2.4.1).

2.4 Power Transfer Through Free Space Using Electromagnetic Wave

In electromagnetic radiation mode, electric energy is converted into electromagnetic energy for transmission that can be radiated into space. The radiated energy travels in the form of electromagnetic wave that can be received at some other distant point in the space. The electromagnetic wave is then rececived using an electromagnetic receiver where it is converted back into electric energy to reproduce the information contained in the electric signal.

2.4.1 Application of Steinheil's Earth Return Path for Signal Transmission in Free Space

The electromagnetic radiation is a form of energy that propagates through free space or through material medium. It consists of both oscillating electric and magnetic fields in the planes perpendicular to each other and to the direction of propagation. Such signals are used as carrier signal for carrying baseband signal through space, and hence are useful in wireless signal transmission for wireless telecommunication systems.

Wireless communication required communication between two distant points using electromagnetic waves travelling through space as a forward path, however, there must be a return path between these two points. In _1838_, Steinheil put forward an idea of the use of earth as return path as a part of circuit to be used in wireless communication and the concept of return path is shown in Fig. 2.6.

Fig. 2.6 Steinheil's idea of return path for wireless signals

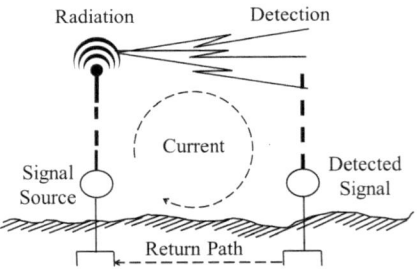

2.5 Conclusion

This chapter investigates the importance of Benjamin Franklin's experiment about investigation of electric spark generated by lightning and its role in the invention of electromagnetic waves which plays a crucial role even in today's wireless radio communication systems.

Chapter 3
Electromagnetic Waves in Communication

Abstract The electromagnetic waves consist of two components of electric and magnetic fields and André-Marie Ampere showed that a wire carrying current acts like a magnet, while Michael Faraday started thinking the other way that when current carrying wires can deflect magnetic needle, then why magnets cannot produce electric current? Sir Willian Thomson theoretically estimated the actual induced current in a wire under the influence of magnetic body in the state of relative motion. It was apparently clear that electromagnetic waves cannot be studied in isolation, and hence interaction between electric and magnetic has to be studied without giving the importance to static surrounding conditions like electrostatic and electromagnetic phenomenon individually, which was rightly covered by M. W. Weber's theory. James Clark Maxwell understood the importance of importance to dynamic surrounding conditions of such bodies holding electric and magnetic properties to derive the theory which he called as "Dynamical Theory" that matter in motion in the space producing electromagnetic phenomenon. However, James Clark Maxwell could not establish the practical existence of such waves and only after 1887, German physicist Heinrich Rudolf Hertz demonstrated the practical existence of electromagnetic waves. It was in 1892 when William Crookes predicted that electromagnetic waves can be used for wireless telegraphy. The roots of electromagnetic radiation detection trace back to the experimentation by Pierre Guitard in 1850 when he observed that when dusty air is electrified, these particles come together or cohere to get lined up in the form of string to exhibit a phenomenon called coherence. Early coherers remained latched up after detection and required mechanical tapping of coherer devices and it was only after April 27, 1899 when Jagadish Chandra Bose invented "Self Recovering Coherers" that worked without hindrance. Meanwhile, many inventors worked on cohering devices.

© The Author(s), under exclusive license to Springer Nature Singapore Pte Ltd. 2021
V. Patil, *Chronological Developments of Wireless Radio Systems before World War II*,
https://doi.org/10.1007/978-981-33-4905-6_3

3.1 Theoretical and Practical Research on Electromagnetic Waves

In *1819*, Hans Christian Oersted [29, 30] observed that when the magnetic needle is placed near the wire carrying electric current, produced a deflection in it that is perpendicular to the direction of electric current (Fig. 3.1a) and when current is reversed, needle jerked to 180°. In *1820*, André-Marie Ampere [31] shown that a wire carrying current acts like a magnet and demonstrated the phenomenon of attraction and repulsion between two different wires carrying the currents.

In *1810*, Michael Faraday joined the London city Philosophical Society promoting intellectual debates and lectures on the topics of current interests for the purpose of self improvements, and due to his active participation in debates, he was given the ticket to attend the lecture of Humphry Davy at Royal Institute in *1813*, Davy hired Faraday to work as a laboratory Assistant at Royal Institute. While working with Davy, Faraday was exposed to the contacts of great inventors and scientists. Faraday got motivated by the work of Oersted and started a series of experiments in *1821* that led to a discovery of Electromagnetic rotation [32] and published his results in a quarterly journal of science, *October, 1821*.

In *1824*, Michael Faraday, started thinking the other way that when current carrying wires can deflect magnetic needle, then why magnets cannot produce electric current? He excited one wire to see that if current can be produced in the other. His experiment was failed, when current is switched on one wire other wire experienced only momentary current and no sustained current occurred, indicated that momentary induction in other wire was caused by the sudden change of current in source wire. Davy guessed the problem and made a right conjuncture that it is not merely sufficient that magnetic field is present, but also requires changing magnetic field to get the changing current induced in the wire. His conjuncture was then demonstrated by showing that a coiled wire produces a current when the magnet moves in and out through the coil. Later on, two years after the death of Davy in *1831*, Faraday conducted exhaustive experiments on electromagnetic induction (Fig. 3.1b) and was able to show that mechanical energy can be converted into an electrical energy.

Fig. 3.1 Oersted and Faraday's experiments. **a** Oersted's experiment. **b** Faraday's experiment

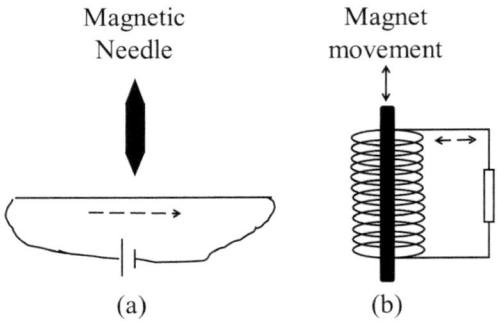

Magnetic Needle

Magnet movement

(a)   (b)

In *1833*, Faraday became Fullerian professor at Royal Institute, in *September 13, 1845*, Faraday conducted an experiment in which he placed the glass over the poles of an electromagnet and passed polarised light through the glass, and when he turned the electromagnet, the polarisation of light changed and he called this phenomenon as "Magneto-optical Effect or Faraday Effect" [33, 34]. From this experiment, he concluded that the polarised light (light wave travelling in one plane) travelling through glass is influenced by the magnetic force, and in turn, glass could be affected by the magnetic force. He called this effect as "Diamagnetism". This work of Faraday showed that electromagnetism and light are related.

The Faraday's work laid the foundations of electromagnetic waves that was further exploited by Sir Willian Thomson (Lord Kevin) and James Clerk Maxwell. Sir Willian Thomson [35] theoretically estimated the actual induced current in a wire under the influence of magnetic body in the state of relative motion. The induced current is given by Eq. (3.1).

$$\because \ \ idU = ki^2dt \ \ \ \therefore \ \ i = \frac{1}{k}\frac{dU}{dt} \tag{3.1}$$

where i denotes the intensity of current that exists in closed conductor, and the amount of work lost by the existence of such current in a loop during time dt is so small that i remains constant, and k is constant depending on the resistance of the wire. U is a forcing function that changes from U to $U + dU$ during time dt.

The Faraday's work brought great inspiration to Edinburgh (UK) born physicist James Clark Maxwell [36]. He studied the mechanical phenomenon related to dynamic bodies based on electric and magnetic conditions of these bodies. Thus, it is a study of motion of these bodies having certain states that how these bodies set in motion by the mutual action of both electric and magnetic conditions. It is the study of statistical mathematical theories of electricity and magnetism of mechanical action between conductors carrying the currents and conductors in which currents are induced. The force between two bodies is considered to be dependent on state/condition of bodies and relative position without giving due importance to surrounding conditions like electrostatic and electromagnetic phenomenon, which was rightly covered by M. W. Weber's theory.

3.1.1 Maxwell's Electromagnetic Field Theory

In this connection, James Clerk Maxwell proposed a theory in *1888* called "Electromagnetic Field Theory" that gives due importance to the space around electric and magnetic bodies. Maxwell says this theory can also be called as "Dynamical Theory" because matter in motion in the space produces electromagnetic phenomenon. The bodies or matter with electric or magnetic conditions is surrounded by an electromagnetic field that is a part of its space around them. The space can be filled with any type of matter that may be just gross matter or empty space or vacuum contained in a

Geissler Tube invented by the German physicist and glass-blower, Heinrich Geissler in *1857*, also called a Vacua. Although Geissler Tube (Vacua) contains vacuum, still contain enough matter to transmit undulations of light and heat across its electrodes and transmission across the electrodes is not greatly changed even though small amount transparent bodies of rare or inert gases are added, still so called as vacuum. This makes us to believe that space in Geissler Tube in action is filled with an aethe-real medium or a medium containing of light particles or bodies spread across whole space in the tube, these bodies are permeable and capable of setting up the motion to move them from one part to the other called transmission called electromagnetic radiation.The electromagnetic waves are caused by the coupling of changing electrical and magnetic fields, once created, it continues on forever unless it is absorbed by a matter.

Maxwell's proposed a dynamical theory called "Electromagnetic Field Theory" that gives due importance to the space around electric and magnetic bodies, this is because the medium is capable of receiving and storing two types of energies called "Energy based on Motion" and "Potential Energy" that allows medium to do some work for recovering from the displacement by virtue of elasticity.

Although there is a rotation of the plane of polarisation by magnetism, the properties of magnetic fields remain unchanged by a change of the medium. When any bodies is moved through the lines of magnetic force, it experiences a force called electromotive force and the two ends of body get oppositely charged that is responsible for electric current to flow through the body. However, if electromotive force acts on body consisting of dielectric, its parts get polarised and similarly distributed and under the continuous change of magnetic force, polarised parts of material gets aligned accordingly while maintaining the continuality and signifies the response of dielectric body that responds to ac currents. The electromagnetic field theory, intern highlights and explains the great role of field theory in the explanation of electromagnetic waves.

3.2 Electromagnetic Radiation and Related Historical Developments

The curiosity about how lightning phenomenon occurs? was responsible for the investigation of properties of electrostatic discharge that passes through the free space and related radiation it emits, gave a birth to a electromagnetic wave concept and the electromagnetic wave radiated can also be called as electromagnetic radiation.

3.2.1 How Electromagnetic Wave Concept Originated?

The phenomenon that how the electric current passes through any medium like space, increased the curiosity of researchers. Researchers through the observation of various phenomena like how lightning strikes, and why sound or thunder is produced when lightning strikes? Thinking of these aspects gave birth to the concept of electromagnetic waves produced through electric discharge. The electromagnetic waves produced by electric discharge through space have both electric and magnetic fields associated with them. The electromagnetic wave induce a voltage in copper wire coil or when the coil is supplied with the electric current can produce a magnetic field.

3.2.2 Theoretical Studies of Electromagnetic Wave Concepts by James Clerk Maxwell

In _1864_, James Clerk Maxwell theoretically studied the properties of electromagnetic waves [36, 37] and developed the laws for predicting the behaviour of electromagnetic waves in terms of electric and magnetic fields. However, practical detection of electrostatic waves still remained a challenge.

3.2.3 Experimental Demonstration of Existence Electromagnetic Waves by Heinrich Rudolf Hertz

Based on the theory of electromagnetism put forward by James Clerk Maxwell in _1864_, German physicist Heinrich Hertz [38] demonstrated existence of radio waves also called as electromagnetic waves in _1887_. Heinrich Hertz forced electric spark to occur through a gap of dipole the radiated power that was received by the loop antenna which was a classic case of wireless signal transmission through the atmosphere Fig. 3.2.

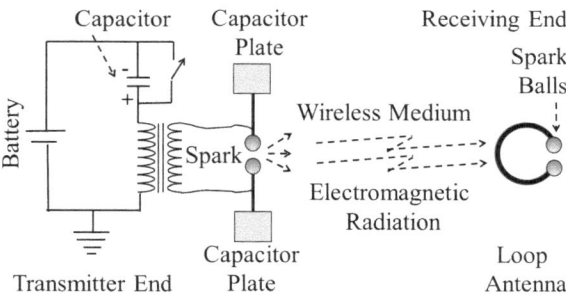

Fig. 3.2 Hertz's experiment of wireless transmission

Heinrich Rudolf Hertz, however, could not realise that his invention will have such great potential in future communications and left its commercial exploitation to others.

3.2.4 Establishment of International Telegraph Union

The foundation of International Telegraph Union (ITU) was laid in International Telegraph Convention at Paris when twenty founding member countries signed a document to establish International Telegraph Union (ITU) on *May 17, 1865* for the purpose of managing wired International Telegraph Networks and International Telegraph Union (ITU) was born. International Telegraph Conference of 1868 held in Vienna decided that ITU will operate from its own bureau in Berne, Switzerland.

With the passage of time, telegraph networks became wireless and after the invention of voice telephony its mandate was changed to cater the changing needs of wireless radio telephony. The International Telegraph Conference of Berlin in 1885 proposed an international legislation governing telephony.

3.2.5 Applications of Electromagnetic Waves in Real Life by Nicola Tesla

In *1881*, Nicola Tesla after moving to Budapest, Hungary to work with Tivadar Puskàs at Budapest Telephone Exchange, where he perfected repeater or amplifier. In *1884*, he moved to the US and in late *1886*, he met Alfred S. Brown a superintendent of Western Union who helped him to promote his work for financial gain. Alfred S. Brown helped Tesla in negotiating his invention of polyphase induction motor with George Westinghouse $60,000 in cash and stock and a royalty of $2.50 per AC horsepower produced by each motor. This gave him time and money to pursue his interests.

In *1886*, Tesla travelled to Paris where he could know the experiments of Hertz and found them of great importance and decided to explore more in this area. From *1890*, Nicola Tesla [39] concentrated to systematically investigate high frequency electric waves with an intension to develop the system for transmission and reception of electrical energy without wires and in *1891*, Tesla demonstrated his high frequency machine [40, 41] before American Institute of Electrical Engineers at Columbia college, It is basically tunable circuit for a given frequency for maximum energy transfer using coherer (a primitive from of radio signal detector). The Tesla's high frequency machine is shown in Fig. 3.3.

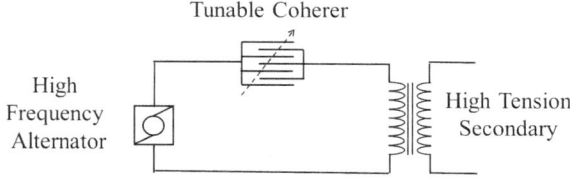

Fig. 3.3 Tesla's high frequency machine

3.2.6 William Crookes Prediction About Use of Electromagnetic Waves for Wireless Telegraphy

Meanwhile, in *1892*, William Crookes predicted that electromagnetic waves can be used for wireless telegraphy, this was a time that even existence of electron was not predicted. Five years later, Joseph John Thompson announced the existence of electrons.

3.2.7 Nicola Tesla's Explanation of Wireless Transmission Concept at Franklin Institute

In *March 1893*, Tesla explained his wireless transmission concept at Franklin Institute, in Philadelphia, PA, USA and National Electric Light Association, St. Louis, MO USA and covered a topic called "On Electrical Resonance", and how electrical energy can be transmitted through earth as shown in Fig. 3.4. The high tension generator transmitter is connected to earth and elevated capacity to impress oscillations on the earth. A distant receiver also connected to earth and elevated capacity to collect energy to activate suitable device.

Tesla got a patent (US Patent No. 568178) for his tuned circuits in *September 22, 1896* and transmitting and receiving circuits (US Patent No. 613809) on *November 8, 1898* as shown in Fig. 3.5.

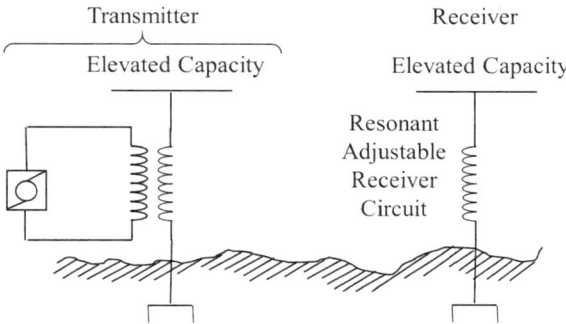

Fig. 3.4 Electric energy transmission through earth

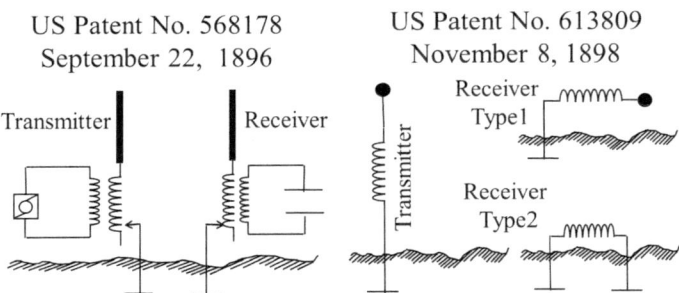

Fig. 3.5 Patents of Nicola Tesla

3.2.8 Radio Frequency Detection and Coherence Phenomenon

The radio frequency generation and detection were the most important aspects in early wireless radio communication systems. In *1842*, Professor Joseph Henry while experimenting with Leyden jar, noticed the discharge of Leyden jar in the form of a capacitor that magnetised the needles lying in the basement. The roots of radio frequency detection trace back to the experimentation by Pierre Guitard in *1850* [42]. Guitard observed that when dusty air is electrified, these particles come together or cohere to get lined up in the form of string to exhibit a phenomenon called as coherence and device responsible to exhibit this phenomenon is called as "Coherer". Others who studied this phenomenon [43] were Samuel Alfred Varley; Edwin J. Houston and Elihu Thomson; David Edward Hughes; Temistocle Calzecchi-Onesti.

3.2.9 Applications of Induction Coils and Their Roles in Spark Generation

Samuel Alfred Varley [44] in *1866* patented dynamo on *December 24, 1866*, Professor Edwin J. Houston working with his assistant Elihu Thomson [45] published a paper "On new connection for the Induction Coil" in *1971* issue of Journal of the Franklin Institute.

In *1879*, Professor David Edward Hughes [46] worked on a mechanism that produced the series of spark by interrupting induction balance using clock mechanism and the aerial waves produced due to spark were picked up at the range of 500 yards in his telephone device. On *February 20, 1880*, his work was presented before the Royal Society.

In *1884*, Italian physicist Temistocle Calzecchi-Onesti [11] worked on spark gaps but all these above described inventors they did not use it.

Fig. 3.6 Édouard Eugéne
Désiré Branly's Coherer

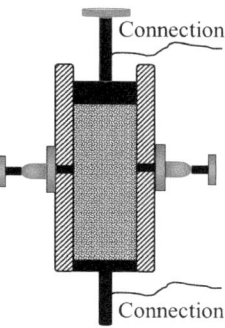

In *1888*, Oliver Joseph Lodge, sparked the debate "Practice versus Theory" [47] before "The Society of Art" by stating that self inductance of wire is more important than the resistance when lightning strikes it. The claim of Oliver Lodge was in a direct contradiction with the recommendations made in Lightning rod conference of *1882*. To prove his point, Oliver Lodge simulated lightning using the discharge of Leyden's jar while Preece and others questioned the validity of an analogy put forward by Lodge. The main issue was that for self induction in radically increasing impendence was due to rapidly changing alternating current and according to Maxwell theory as explained by Heaviside, Poynting and John William Strutt (Lord Rayleigh), it was the field effect responsible for electric current.

The real twist of its use came in the year *1891* when French physician Édouard Eugéne Désiré Branly showed that magnetised particles placed within the field have tendency to cohere similar to attraction of iron particles by the magnet in which iron particles get magnetised in the direction magnetic field and his coherer is shown in Fig. 3.6.

3.2.10 Elihu Thomson's Spark Gap and Aerial Circuits for Wireless Transmission

In *1892*, Elihu Thomson, experimented oscillatory aerial circuits using transformer for isolating two circuits, primary circuit consisted of spark gap and capacitor while secondary was tuned to raise the potential which was published in Electric World in *February 20* and *27, 1892* which was later used by Tesla in *1893*.

3.2.11 Role of "Coherer" in Wireless Radio Communication

The coherer was considered to be an important device in wireless radio communication. The term "Coherer" was coined by Sir Oliver Joseph Lodge [48] in *1893* and in *1894*, he used better technique than Édouard Branly and constructed a coherer using iron filling by adding a mechanism for quaking (shake with fast, tremulous

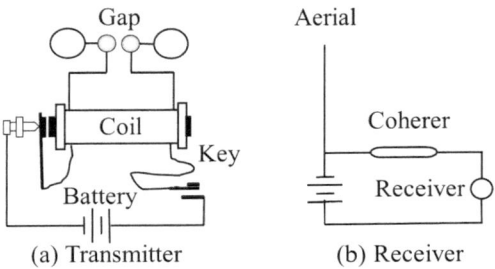

Fig. 3.7 Electric telegraphy circuits of Oliver Joseph lodge

movements like seismic vibrations) and trembling (vibrating slightly and irregularly) to bring it into its original form that was latched into a different state after its excitation. The coherer resistance drops sharply when ac or dc voltage beyond a certain threshold is applied to it and this sharp change of resistance is from tens of thousands to tens of ohms used for RF signal detection. A coherer when used together with RF circuit and battery in series, RF signal from an antenna, exceed the threshold of coherer that lowers its resistance from the order of 10^4 ohms to few tens of ohms, that allows current to flow through coherer. The signal part received from an antenna is then separated. Oliver Joseph lodge obtained a US patent No 609154 for his apparatus on *August 16, 1898* that describes the use of coherer [49] in receiver as shown in Fig. 3.7b while transmitter part is shown in Fig. 3.7a.

3.2.12 Increasing Antenna Capacity

To increase the capacity of antennas at both transmitter and receiver ends, Lodge used large areas in the form of conical shape and large conductors in the form of coils across spark gap on transmitter side and at receiving end to collect the signals. He distinguished long distance transmission from others and used different arrangements. The Fig. 3.8 shows capacity improvements for long distance communication between transmitter and receiver.

3.2.13 Augusto Righi's Oscillator

In *1895*, Augusto Righi of Bologna invented spark gap also called as "Righi Oscillator" [50] that was used by Guglielmo Marconi in his transmitter circuit that was demonstrated to the Post Office, in his historic demonstration held at three mile hill on Salisbury Plain on *September 2, 1896* using parabolic reflectors. Every time the Morse key is pressed, spark is created and information is transmitted. The three spark gaps created in Righi Oscillator helps in creating more strength in the transmitted signal. The Righi's Spark Gap is shown in Fig. 3.9.

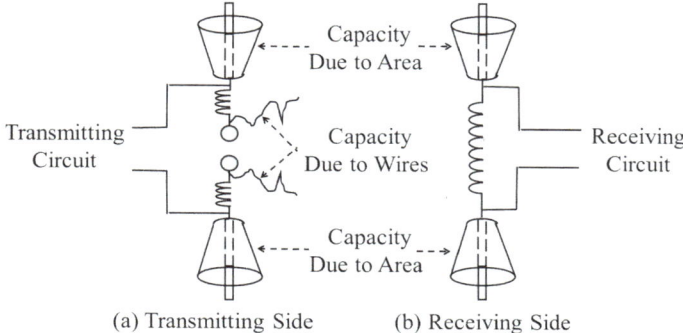

Fig. 3.8 Capacity improvements of transmitting and receiving circuits by Oliver Joseph Lodge

Fig. 3.9 Righi's oscillator or Spark Gap, *Credit* CC BY 4.0, John J Griffin & Sons Ltd, Righi pattern oil-bath oscillator, |1910|, Science Museum Group. Righi pattern oscillator, 1870–1923. 1923-444 Science Museum Group Collection Online. Accessed July 23, 2020

3.2.14 William Gladson's Experimental Wireless Transmitter

In *1897*, William Gladson [51, 52] at University of Arkansas, Fayetteville, Arkansas, reported to have developed experimental wireless transmitter and in *1900*, a wireless telegraph station was erected at University of Arkansas, which become land based licensed station with call sign 5YM in *Dec 04, 1923* and further it became KFMQ medium wave broadcasting station in *January 1924*.

3.2.15 Eugene Ducretet's Experimental Wireless Transmission from Eiffel Tower

The high rise structure like the Eiffel tower was an attraction for wireless equipment experimentation and even before its completion, in *November 1897*, Eugene Ducretet experimented his wireless transmission equipment from the third floor of Eiffel tower,

Fig. 3.10 Coherer
reactivation using
mechanically tapping
mechanism operated by an
electromagnetic coil, *Credit*
CC BY-SA 3.0 |1917| H.
Winfield Secor, Radio
Detector Development,
Electrical Experimenter,
January, 1917

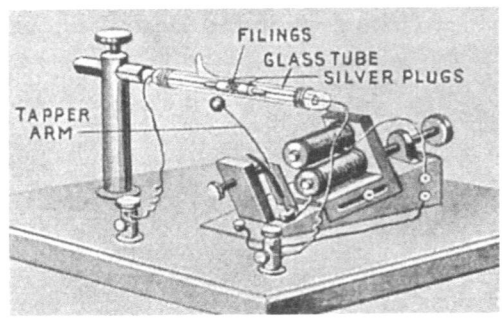

and further, on *November 5, 1898*, he made long wave communication to Pantheon in Paris, at the distance of three miles using his spark transmitter.

3.2.16 Karl Ferdinand Braun's Loosely Coupled Spark Gap Oscillator

In *1899*, Karl Ferdinand Braun, A German physicist, realised that a circuit with loose coupling between spark gap oscillator and antenna circuit produces better results in terms of less damped oscillations and radiated more energy from the antenna that was distributed over a narrower band of frequencies. He filed a British patent for his work in *January 26, 1899*. The principle RF detection using crystal-based device was discovered in *1874* by German researcher and Nobel Laureate Karl Ferdinand Braun [53, 54] (who received Nobel Prize in *1909*). His work on crystal diode detector was patented in *1899*.

Early coherers once cohered remained cohered and has to be decohered by mechanical tapping posing problems in continuous reception of radio signals and every time signal is received, required mechanical hammer operated by electrical relay and arrangement was only suitable for telegraphy where distinct signals related to dash and dots and not suitable for audio signals used in telephony. The mechanical tapping mechanism used to restart coherer is shown in Fig. 3.10.

3.2.17 J. C. Bose's Self Recovering Coherer

Sir J. C. Bose, a professor of physical science started working towards "Self Recovering Coherer". In *1895*, he started systematic study [55, 56] regarding cohering properties of different materials to be used in coherer and experimented with various types of materials for coherers and came out with U shaped self recovering coherer [57, 58] as shown in Fig. 3.11. These self recovering coherers were constructed using metals and semiconductors contacts. This invention was communicated by

Fig. 3.11 Self-regulating coherer of Sir J. C. Bose

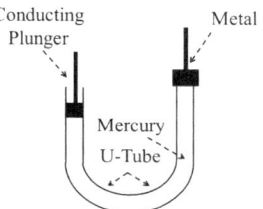

Lord Rayleigh, FRS, that was received by Royal Society on *March 6, 1899*, read on *April 27, 1899*.

Lieutenant L. Solari started experimenting on self-regulating coherer of Sir J. C. Bose in *February 1901* and presented his trivially modified version to Marconi [59], who applied for a British patent and got a patent No. 18105. The patented device used straight tube coherer instead of U tube coherer that was used in transatlantic wireless transmission by Marconi on *December 12, 1901* was similar to J. C. Bose's mercury coherer.

3.3 Conclusion

In conclusion, this chapter deals with electromagnetic waves radiated due to the generation of sparks provided great deal of interest amongst many inventors and quest for knowing more about electromagnetic waves led to theoretic investigation by James Clark Maxwell while after practical detection of electromagnetic waves by Heinrich Hertz, these waves were called as "Hertz's Waves". After this practical demonstration, many inventors lined up for how to continuously generate and detect these radiated waves to be fruitfully used in wireless communication.

Chapter 4
Marconi's Exploitation of EM-Waves

Abstract Marconi's experimentation started with the extension of physical ranges of wireless transmitters based on spark energy applied coded messages based on Morse code. Marconi's initial experimentation was carried out in his father's country estate house "Villa Griffone" at Pontecchio, Bologna, Italy. His first success was based on the "Experiment of Celestin Hill" He was drastically progressed when he moved to London where he came in contact with William Preece, chief engineer of the British post office who provided a great platform for his efforts and tried to connect land mass to ships, island to island and ship to ship using wireless communication. With these experimental successes, he dared to carry out the successful wireless transmissions over the Atlantic ocean covering a distance of 3,000 kms (1865 miles). It was the first time for Marconi when official messages were transmitted and commercial messaging between Britain and Canada across the Atlantic started. For his efforts, he was awarded the Nobel prize in 1909 which also was shared by Ferdinand Braun for his contribution to wireless telegraphy in the form of a crystal detector.

4.1 Marconi's Work on Wireless Transmission

In *1894*, Guglielmo Marconi started pursuing his goal towards building long distance wireless transmitter and started re-inventing the work of personalities like Maxwell, Hertz and Righi. He tried to emulate a spark transmitter built by Righi that was very similar to that of Hertz. It used two Morse Keys for coding in the form of dash and dots of Morse signalling for transmission of text information that can be understood by a skilled person. The coherer was activated using radio signals that triggered relay action to activate the bell. The experiment was performed in an attic room at his father's country estate house "Villa Griffone" at Pontecchio, Bologna, Italy. The photograph of this experiment [60] is shown in Fig. 4.1.

After the preliminary success, Marconi started trying his wireless transmission for various ranges in his house and was successful in sending signals to the ground floor of four storeyed house, and in next try, he could send the signals at the end of the garden. After the progressive successes of Marconi, his father acknowledged

© The Author(s), under exclusive license to Springer Nature Singapore Pte Ltd. 2021   33
V. Patil, *Chronological Developments of Wireless Radio Systems before World War II*,
https://doi.org/10.1007/978-981-33-4905-6_4

Fig. 4.1 Photograph of Marconi's early experiment. *Source* CC BY-SA 3.0, Guglielmo Marconi, |1926| looking back over thirty years of radio, radio broadcast magazine, Doubleday, Page, and Co., Garden City, New York, Vol. 10, No. 1, November 1926, p. 31, American Radio History

his efforts and agreed to financial support his experimentation. In September *1895*, he conducted outdoor experiment called "Experiment of Celestin Hill", he could successfully send the signal to this hill that was around 1.8 miles away from his house. The antenna was made of heightened structure made up of metallic rods that could make a great difference in his success. The Marconi's coherer consists of tapered silver conductors placed within an evacuated glass tube, and used in his wireless transmission system [25, 42, 61] that was more or less similar to Hertz's experiment that sent Morse code to a couple of miles. The details of this experiment are shown in Fig. 4.2.

On *12th February 1896*, Marconi moved to London on the initiatives of his cousin Henry Jameson Davis and under his guidance Marconi filed a British Patent No. 12039 on *June 2, 1896*. In London Marconi, met William Preece, chief engineer of the British post office and he demonstrated his apparatus to Post Office, this historic demonstration was held at three mile hill on Salisbury plain on *September 2, 1896* where post office, navy and army officers were present and the results were not convincing and merely a range of a half a kilometer was achieved and with parabolic reflectors range was increased to 2.5 kms which was not still convincing as per the expectation. Yet another demonstration was held within six months at Salisbury plain on *March 1897*, in which aerials of 40 m were hoised using kites and balloons that increased the range by a factor of two. At Salisbury Plain, many more demonstrations were continued even in *1897* and covered a range up to 11 kms.

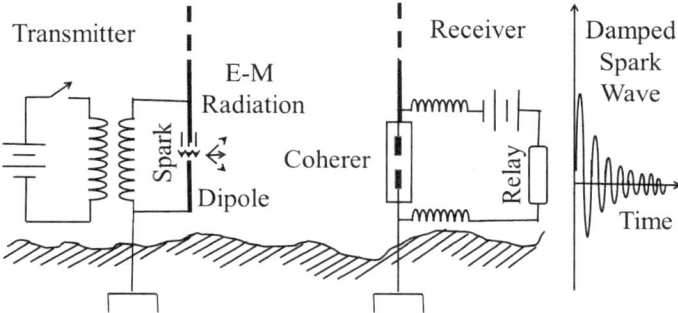

Fig. 4.2 Credit: CC BY-SA 3.0, Guglielmo Marconi, looking back over thirty years of radio, radio broadcast magazine, Doubleday, Page, and Co., New York, Vol. 10, No. 1, November 1926, p. 31 |1926|, American Radio History

Fig. 4.3 A demonstration at Flat Holm Island on May 13, 1897. *Source* Marconis first wireless message transmitted over the sea, from Flat Holm to Lavernock Point, Cardiff Council Flat Holm Project, May 13, 2016, https://wearecardiff.co.uk/2016/05/13/

4.1.1 Marconi's Demonstration at Island in Bristol Channel

Marconi also held the demonstration was held at Flat Holm, a limestone island lying in the Bristol channel in *May 13, 1897*, and after some unsuccessful attempts due to bad weather, he successfully transmitted signals to a range from 13 to 15 kms using aerial of 92 m and 20 in. spark coil. The demonstration set is shown in Fig. 4.3.

4.1.2 Marconi's Experiments at South East Coast of England

An Island and County in the South coast of England called "Isle of Wight" was the sight of Marconi's experiments. In *1897*, he experimented communication with ferry boats from his apparatus installed at Royal Needles Hotel, Alum Bay. He hired two ferry boats and successfully communicated with mainland base station at Madeira

(a)   (b)

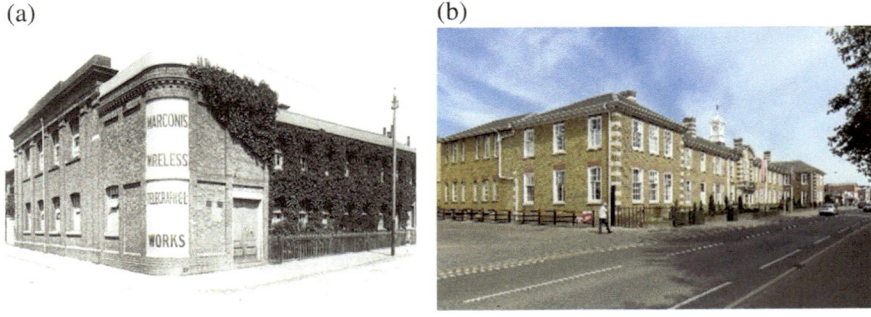

Fig. 4.4 The Marconi Factory site the old and new **a** World's first radio factory established by Marconi in December 1898, Credit: Marconi Heritage Group and **b** The recent look of former Marconi Factory building, Chelmsford, Credit: CC BY-SA 2.0, ©John Sutton—Geograph.org.uk/p/5818432

Fig. 4.5 Marconi's modified circuit of receiver

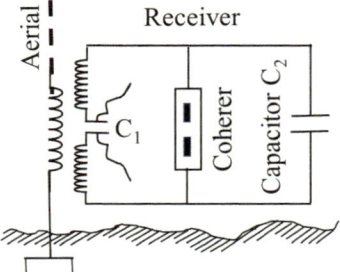

House in Bournemouth. He also demonstrated his equipments in Roma and the Gulf of La Spezia, *July 1897*, covering a range of 16 kms.

In the same year *July 20, 1897*, Marconi registered world's first radio factory by name "The Wireless Telegraph and Signal Company Ltd." that was setup in an old silk factory premises on *December 1898*, at Hall Street, Chelmsford, Essex, England (Fig. 4.4). Its name was changed to "Marconi's Wireless Telegraph Company Ltd" in *1900*, through which he was successful in commercialising wireless radio systems.

From the beginning of *1898*, the circuit of receiver shown in Fig. 4.2 was abandoned [62] and instead of connecting aerial directly to coherer, it is connected as shown in Fig. 4.5. This arrangement allowed a certain degree of adjustment of tuning between the transmitting end and receiving end by a varying period of oscillations of transmitting antennae.

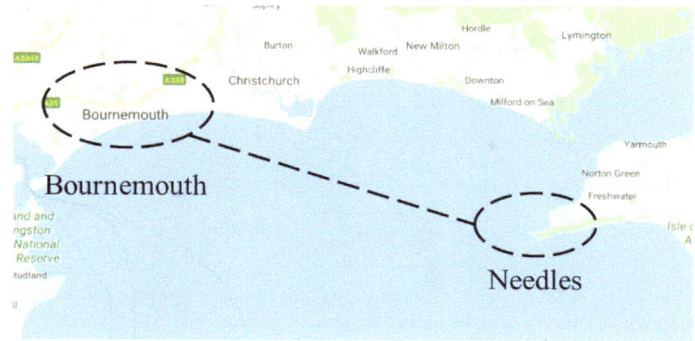

Fig. 4.6 Marconi's first commercial link between Needles and Bournemouth. Courtesy Google Maps

4.1.3   Marconi's First Public Commercial Radio Telephony Service Between Needles and Bournemouth

On *June 03, 1898*, Marconi opened the first public commercial radio telephony service between Needles and Bournemouth covering a distance of 30 kms (Fig. 4.6) and also provided radio telegraphy service to lighthouse, commission of Llyod, Journalistic services, royal yacht and so on. In the summer of *1898*, Queen Victoria and Marconi happened to be on Isle of Wight, the queen at Osborne house and Marconi at his shore station near Needles, where he got special communication from the Queen to provide a wireless link between herself and her son who had a knee injury and traveling on board the Royal Yacht during Cowes Week. She wanted to frequently inquire about the wellbeing of her son through wireless messaging and the first of 150 messages was on *August 04, 1898*. During Christmas Eve of *December 24, 1898*, ship to shore communication took place between Foreland Lighthouse, near Dover and Goodwin Lightship in the English Channel.

4.1.4   Marconi's Station at South Foreland

At the end of *1898*, Marconi setup the station at South Foreland and the equipments installed were more powerful than planned earlier to communicate with East Goodwin lightship at a distance of 12 miles (19 kms).

In early *1899*, Marconi also modified his transmitter [62] by increasing the capacity of transmitting antenna by means of increasing energy storing capacity as shown in Fig. 4.7.

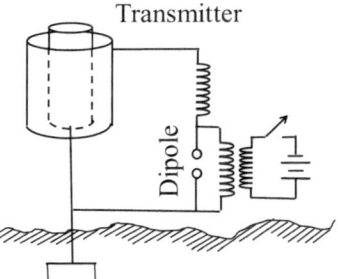

Fig. 4.7 Marconi's modified circuit of transmitter

4.1.5 Marconi's Transmissions over the English Channel

The transmissions over the English channel between South Foreland, Dover, Kent in England and Wimereux in France, covering the distance of 51 kms was established on _March 27, 1899_, along with transmissions between Wimereux-Harwich and Chelmsford (130 kms). The first distress call for assistance was transmitted to a wireless station at South Foreland in _April 1899_ [63], when East Goodwin lightship, marking the south east coast of England was rammed in dense fog by the SS R. F. Matthews and in response, the assistance was provided. Royal Navy exercises _July 1899_ (140 kms) were performed when Marconi was asked by Royal Navy to equip its three ships. In _Oct 1899_, Marconi travelled to United States to equip US warships.

On _April 26, 1900_, he patented the mechanism for prevention of mutual interference between signals (Patent No. 7777) that was inspired by the experiment of Oliver Lodge in _1897_. Fleming who was appointed as consultant in _1899_, signed a contract in _December 1900_ to became a scientific advisor. In _1900_, Marconi constructed a complete system of transmitter and receiver using his usual method of elevated capacity area and earth that was inductively coupled to oscillation circuit consisting of condenser, inductance and spark gap (detector) as shown in Fig. 4.8.

The same way receiver circuits for was developed which are shown in Fig. 4.9.

4.1.6 Marconi's Wireless Transmissions Experiment Between St. Catherines Point on Isle of Wight and Lizard in Cornwall

In _January 1901_, an experiment was conducted between two stations having a distance of 186 miles between them called St. Catherines point on Isle of Wight and Lizard in Cornwall and was successful (Fig. 4.10). The height of these stations at both ends had a height not more than 100 m but to clear the curvature of the earth, the required height was estimated to be 1600 m, however, the results showed that

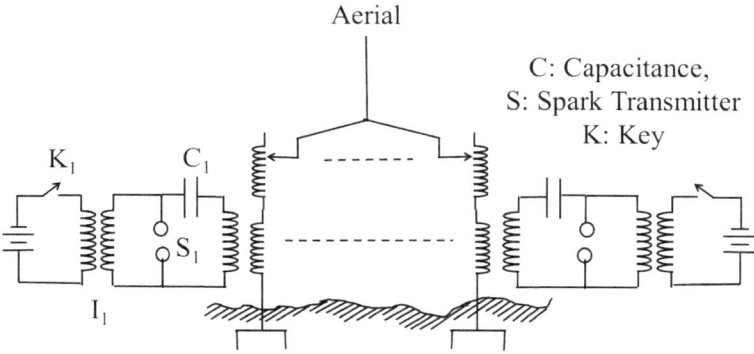

Fig. 4.8 Marconi's circuit of transmitter for multiple differently tuned circuits

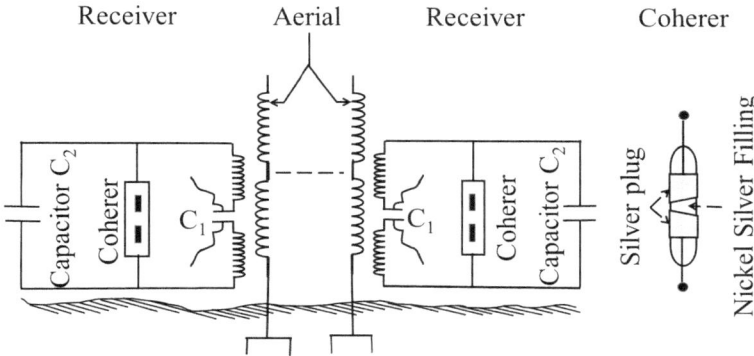

Fig. 4.9 Marconi's circuit of receiver for multiple differently tuned circuits

such heights were not required. Encouraged by such results, he concluded that any distance can be covered with the belief that earth's curvature does not stop the propagation of waves and set his targets to very large distances and dreamed of connecting Europe with America without wires to cross the Atlantic ocean.

4.1.7 Marconi's Wireless Transmissions over the Atlantic Ocean

Marconi set a target to cross the Atlantic Ocean and in this connection, he visited South-West Cornwall in *July 1901* and selected a place called Poldhu. The place was situated on the Lizard Peninsula, a small area in south Cornwall, England, UK, a headland site where suitable projected rock mountain highland projecting inside the sea was available. Due to its suitability, the site was selected. In *October, 1901*, site

Fig. 4.10 Marconi's wireless communication link between Needles and Lizard. Courtesy Google Maps

Fig. 4.11 Photograph of Marconi's aerial erected on 17 September, 1901, at Poldhu, *Source* CC BY-SA 3.0, Autor unbekannt (Unknown Author) |December 1901| The Marconi Company first antenna system at Poldhu, Cornwall, December 1901, archive of Marconi Corporation, PLC, 1901

development was started to construct first transatlantic transmitter powered by 25 KW alternator driven by 32BHP engine. A giant aerial [64] was erected (Fig. 4.11), and when work on aerial was on completion on *17 September 1901*, it was collapsed due to strong winds.

However, without losing heart, he again constructed a temporary antenna at Poldhu (*December 1901*) using stout vertical poles used for supporting sails [65] as shown in Fig. 4.12. The wires were arranged such that it formed a conical structure with one end of wires converged to one point and other ends were hanged on horizontal support using insulated stay between masts having a distance of 60 m between them and height of 48 m.

A station built at other end called Cape Cod, a hook-shaped peninsula in Massachusetts, USA which was any way destroyed in storm, and hence Marconi decided to change its location to St. John's, Newfoundland Island, (now part of Canada). At Newfoundland, aerials were raised up to 600 feet using six kites and two hydrogen balloons filled using 25 cylinders of gas as shown in Fig. 4.13.

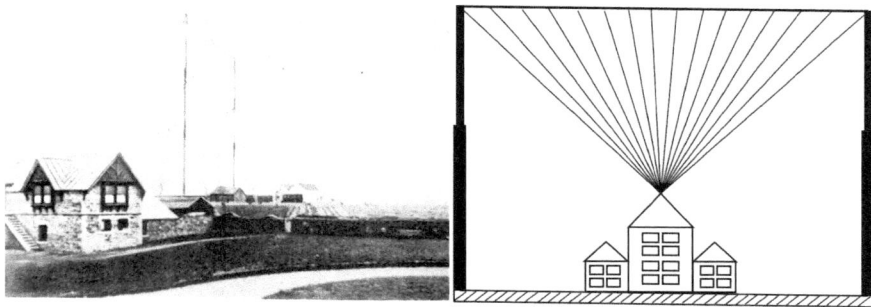

Fig. 4.12 Marconi's reconstructed aerial after the collapse of earlier aerial at Poldhu in December 1901, *Source* Archive of Marconi Corporation, PLC, 1901. Courtesy: Grace Guide, UK

Fig. 4.13 Marconi's aerial at Signal Hill, at Newfoundland receiving station. *Courtesy* Henry M. Bradford, Marconi in Newfoundland: The 1901 transatlantic radio experiment, March 27, 2002 (A painting depicts Marconi's receiving aerial wire supported by a kite)

On *December 12, 1901*, after some initial problems, Marconi tested his transmitter strong enough to send messages across the Atlantic ocean as shown in Fig. 4.14. As planned, Marconi asked Poldhu station to send telegraphic message to transmit a letter S (consisting of 3 dots) daily for three hours. The message was successfully transmitted to a distance of 3,000 kms (1865 miles) from Poldhu to Signal Hill, St. Johns Newfoundland.

Even at this point, Marconi could not explain "why earth's curvature does not affect radio transmission?", but some of his critics correctly predicted that Marconi's transatlantic radio signal was headed to space that was reflected back to receiving station by F layer of ionosphere. The circuit that is used in the transatlantic radio link is shown in Fig. 4.15.

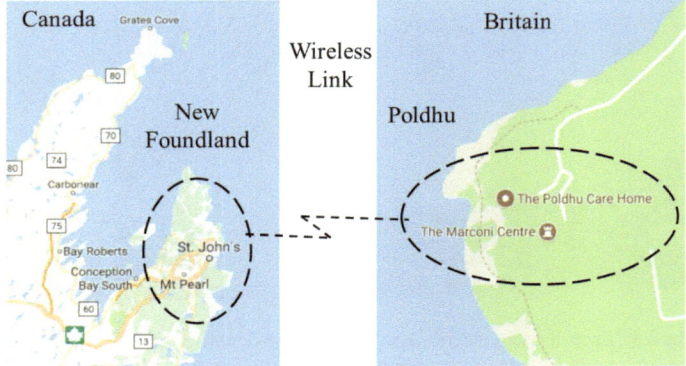

Fig. 4.14 Marconi's transatlantic link December 12, 1901, Courtesy Google Maps

Fig. 4.15 Circuit of
transmitter used by Marconi
at Poldhu in 1901

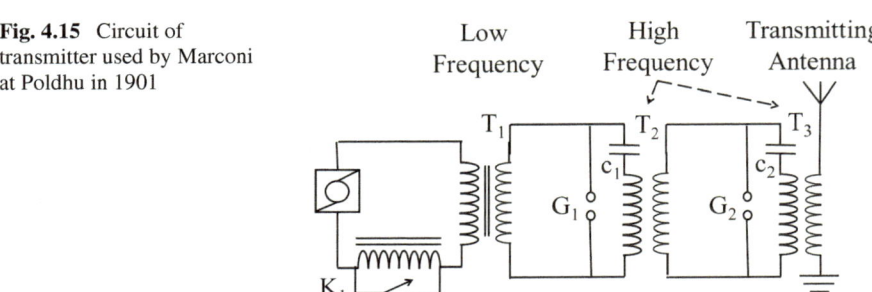

4.1.8 Marconi's Work for Ship Communication

By *1901*, Marconi was operating communication stations on 32 ships belonging to
the British government and also had stations on the Island of Corsica of France in
Tyrrhenian sea on south of France (175 kms from Antibes to Calvi in *April 1901*),
Belgium and Germany [66].

On *February, 1902*, Marconi conducted confirmatory tests between Poldhu and a
receiving stations on ship called SS Philadelphia, an American liner kept at different
distances from Poldhu and readable messages were successfully received by the
instruments on board at the distances 1551 miles and 2099 miles. The magnetic
detector used on board a ship is shown in Fig. 4.16.

It was also observed that the range during night time was double than day time
and Marconi admitted that the phenomenon at that time could not be explained.

In *March 1902*, Nathan B. Stubblefield demonstrates wireless telephony on
Potomac river by establishing communication between from shore of Potomac river
to a steamboat in Potomac river.

During *1902*, he established connection between Poldhu and Italian Carlo Alberta
Cruise that was arranged by the king of Italy.

Fig. 4.16 Magnetic detector
used on board a ship during
1902 summer, *Credit* CC
BY-SA 4.0 Alessandro
Nassiri | 2012 | Museo
Nazionale della Scienza e
della Tecnologia Leonardo
da Vinci

Fig. 4.17 Antenna used at
Glace bay, Nova Scotia
province of Cape Breton
Regional Municipality of
Cape Breton Island, Canada

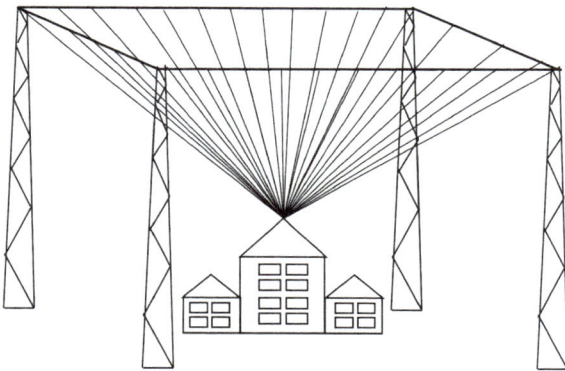

In *December 16, 1902*, Marconi experimented between Poldhu and Glace Bay,
Nova Scotia province of Cape Breton Regional Municipality of Cape Breton Island,
Canada, where one of the antenna structure employed out of various forms and
structures thought by Marconi is shown in Fig. 4.17.

In *January 19, 1903*, yet another American station at Cape Cod, Massachusetts
was established and a message to Poldhu was sent from President Roosevelt to King
Edward VII and in the same year, Italian Ministry of Mail and Telegraph offered
Marconi to build a link from Italy to Argentina with the intension to connect migrated
Italian people. Marconi selected a sight at Coltano [67] in king's estate of San Rossore
near Pisa for setting the facility for this purpose and this sight started in *1904*.

4.2 Berlin International Wireless Telegraph Convention of 1906

In *1903*, a preliminary radio conference was held for discussions to establish international regulations for wireless radio telegraphic communications which were fol-

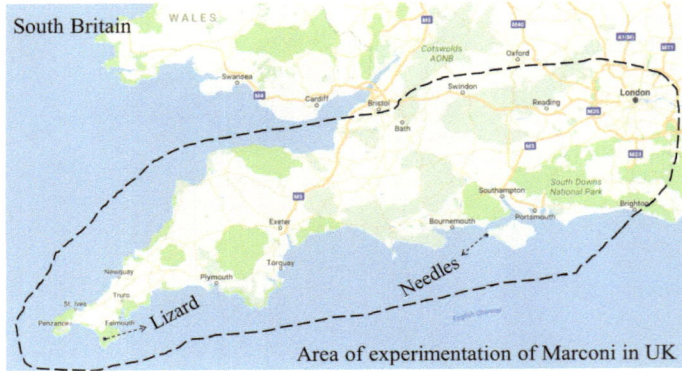

Fig. 4.18 Marconi's experimental area coverage in Britain, Courtesy Google Maps

lowed by another conference called "The first International Radio Telegraph conference [68] at Berlin, Germany on *November 3, 1906*. In this conference, 27 nations signed a number of agreements concerning wireless communication and designation Save Our Souls (SOS) distress calls. Navy Department's Bureau of Equipment of United states brought out the document regarding this convention in *1907*, as a documented evidence.

4.3   Commercialization of Radio Transmitters and Receivers for General Public

In *November 25, 1905*, "Scientific American magazine" carried an advertisement for Electro Importing Company regarding the availability of a pair of battery-operated radio transmitting and receiving devices by "Telimco" called "Telimco Wireless Telegraph Outfit" that guaranteed to work for at least one mile range. It was believed to be the first of its own kind of advertisement offering ready to use transmitting and receiving radio devices for the general public.

It was the first time for Marconi for official messages were transmitted while commercial messaging between Britain and Canada across the Atlantic started on *October 1907*. Marconi's pioneering work [62] earned him a laurels and got Nobel prize in *1909* which was shared by Ferdinand Braun for his contribution to wireless telegraphy.

The area of experimentation of Marconi in Britain was mostly south shore of Britain as shown in Fig. 4.18.

Fig. 4.19 Ferdinand Braun's
semiconducting device using
galena crystal (lead sulfide)
probed using pointed thin
wire. *Source* United States
Early Radio History

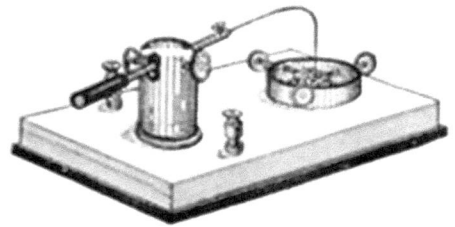

4.4 Braun's Invention of Semiconduction

The invention of crystal detector, also known as "cat's whisker" traces back to *1874*, when young German physicist Ferdinand Braun [53, 69] studied the electrical conduction characteristics of electrolytes and crystals at the University of Wurzberg, Germany. He observed that when galena crystal (lead sulfide) probed using pointed thin wire, conducts electricity in only one direction (a process rectification). This observation opened the doors for converting alternating current (bidirectional directional electric current flow) to direct current (unidirectional electric current flow) used in the detection of radio frequency eaves. This effect, that is semiconduction, was demonstrated by Ferdinand Braun at Leipzig on *November 14, 1876*. The constructional details are shown in Fig. 4.19.

4.5 Exploitation of Semiconducting Principle by J. C. Bose for Wave Detection

The use of this invention (that is, point contacted metallic wire with Galena) for detection of millimeter electromagnetic waves was demonstrated by an Indian physicist Jagadish Chandra Bose [56], a professor of physics at Presidency College in Calcutta, India in *1901*. In *1901*, J. C. Bose applied a patent for point contact semiconductor rectifier for detecting radio signals entitled "Detector for Electrical Disturbances" which was issued a US Patent No. 755840 on *March 29, 1904*. The Jagadish Chandra Bose's crystal detector for millimeter waves and 60GHz apparatus is shown in Fig. 4.20.

4.6 Greenleaf Whittier Pickard's Crystal Detector Patent

On *August 30, 1906*, Greenleaf Whittier Pickard [70] filed a patent for silicon crystal detector that used fine pointed wire for providing delicate contact with mineral providing best semiconductor effect and called as a "cat's whisker and patent No. 836,531 entitled "Means for receiving intelligence communicated by electric waves" was granted on *November 20, 1906*. This revolutionary invention challenged the use

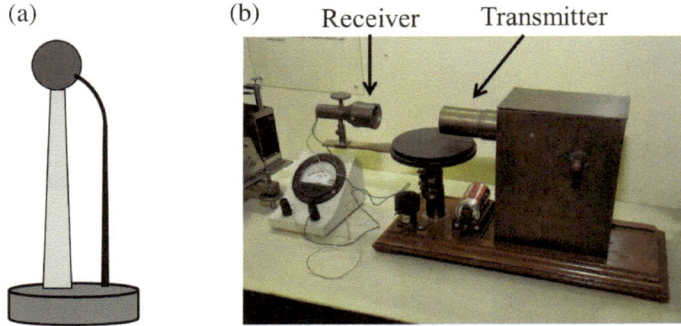

Fig. 4.20 Jagadish Chandra Bose's crystal detector for millimeter waves and 60 GHz microwave apparatus at the Bose Institute, Kolkata, *Credit* CC BY-SA 3.0, Biswarup Ganguly |26 July 2011| Jagadish Chandra Bose Museum—Bose Institute—Kolkata

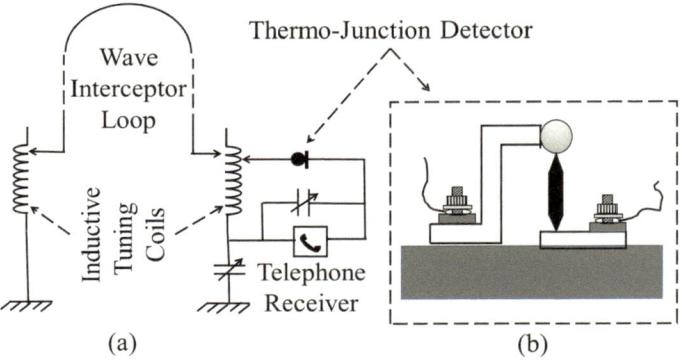

Fig. 4.21 The use and constructional aspects of crystal diode detector

of vacuum diode. The use of crystal detector and constructional aspects as described in his patent is shown in Fig. 4.21.

This "Cat's Whisker detector" was commercially produced and used in early radio receivers from *1906* for a decade. The commercially produced "Cat's Whisker detector" is shown in Fig. 4.22.

4.7 Installation of an Antenna by Northern Europe Wireless Telegraph Service

In *1906*, Northern Europe Wireless Telegraph Service installs its first antenna at Rost, a most wildest and most rugged landscape in Norway, makes history of Nor-

Fig. 4.22 The commercially produced Cat's Whisker Detector, *Credit* CC BY-SA 3.0, Alfred Powell Morgan |1914| wireless telegraph construction for Amateurs, 3rd Ed., p. 134, Fig. 106 on Google Books

Fig. 4.23 Norwegian and European telecommunications' antenna at Rost, Norway, Northern Europe's first fixed wireless telegraph—1906, *Source* Kenth Roger Johnsen, Museumskonsulent, Museum Nord, (Norsk Telemuseum, Srvågen, Lofoten, Norway)

wegian and European telecommunications [71] and remained in focus in the history of telecommunications (Fig. 4.23).

4.8 Conclusion

Marconi was responsible for bringing wireless technology based on spark transmission to the general public, its spread and its commercialization. His efforts would have been incomplete without the ideas, concepts, efforts and inventions put forward by J. C. Bose and Ferdinand Braun.

Chapter 5
Fessenden's Heroic Work in Wireless Telephony

Abstract Earlier wireless systems were based on telegraphic communication in which spark generated was died down after decay period that was active only when symbol was being transmitted while for voice transmission continuous spark was required. At first instance, people tried to increase the duration of spark produced and repetitive production of sparks and "Singing arc" provided the solution to some extent. Elihu Thomson generated continuous train of damped oscillations using circuit consisting of battery, inductor and capacitor. The L and C components were so adjusted that continuous oscillations were produced, for which he got the patent entitled "Method of and Means for Producing Alternating Currents". Danish inventor Valdemar Poulsen successfully demonstrated efficient continuers arc technology that generated higher frequencies. The real breakthrough came in when a Canadian physicist Reginald Aubrey Fessenden started thinking of use of continuous wave in voice transmission and become an important branch of wireless communication called "Wireless Telephony". Initially, Fessenden's efforts were directed to generate continuous sparks for sustained transmission of sound waves using interrupters and transmission was tried at Cobb Island, Maryland, Virginia, on 23 December 1900 and this was the first time in the world that intelligible speech was transmitted using electromagnetic waves. Although Fessenden was successful in transmitting voice continuously using interrupter, he was unhappy with the arrangements of continuous generation of discrete bunches of sparks and hence he invented "Rotary Spark Generation" and further "Synchronous Rotary Spark Generation". Finally, he invented an important heterodyning principle that became a basis for modulation of voice signals over high frequency carrier wave that greatly improved wireless voice communication. General Electric Company's high frequency generators were used to generate continuous high frequency wave to be used as a carrier signal for low frequency baseband voice signals. On Christmas eve of 1906, Reginald A. Fessenden made his historic wireless voice transmission. The US Navy ships fitted with Fessenden's radio sets capable of receiving amplitude modulated (AM) signal and United States Fruit Company ships were notified regarding Christmas eve program broadcast from Brant Rock Station for receiving transmitted signal. This first historic wireless voice transmission paved a way for large scale growth of popular public radio broadcast systems.

© The Author(s), under exclusive license to Springer Nature Singapore Pte Ltd. 2021   49
V. Patil, *Chronological Developments of Wireless Radio Systems before World War II*,
https://doi.org/10.1007/978-981-33-4905-6_5

5.1   The Shift in Research Interests from Wireless Telegraphy to Wireless Telephony

The wireless radio systems developed by Marconi were based on spark technology and he widely used spark transmitters for transmitting the information to the longer distances. The spark gap transmitters had a serious defect that signal transmission could be done only during the fraction of time when spark occurs [72] and hence techniques adopted in spark transmission were only suitable for telegraphic information and were not suitable for continuous signal transmission as required by telephony. By this time it was realised that if telephony signals to be transmitted continuously, spark like signal must be provided contiguously. This idea gave birth to wireless telephony.

5.1.1   Basic Requirements for Wireless Telephony

It was a challenge for maintaining continuity in spark so that audio signal can be continuously transmitted over the electromagnetic carrier generated by the spark signals. Many researchers worked in this direction and developed the devices called "Tickers" or "Ticking Contact" or even called "Tikker". Various types of mechanisms were put forward by numerous researchers. At first instance, people tried to increase the duration of spark produced and repetitive production of sparks. The effective solution was "Singing arc" [73], and at this point distinction should be made between spark and arc, Arc is a continuous portion of some cycle while spark is characteristically represented by momentary flash of the signal.

Elihu Thomson in *1892*, invented the circuit for obtaining a train of damped oscillations as shown in Fig. 5.1a that was excited using DC current. The capacitor and inductances are so adjusted that it oscillates at the frequency decided by L, C and R components and frequencies as high as 40 kHz can be achieved. His work entitled "Method of and Means for Producing Alternating Currents" got patented vide US Patent No. 500630 dated *July 4, 1893* [45]. William Du Bois Duddell used similar circuit except he used resistors in place of inductors as shown in Fig. 5.1b, and produced clearly audible musical note at some distance.

5.1.1.1   Valdemar Poulsen's Efforts to Generate Continuous Spark

In *1903*, Danish inventor Valdemar Poulsen patented his arc generation technology, in which he was successful in increasing frequency and efficiency of arc generation. Unlike discontinuous pulses, Valdemar Poulsen was able to generate continuous arc frequency of 200 kHz that was suitable for amplitude modulation of signals to be transmitted. He patented his invention entitled "Method of producing alternating currents with high number of vibrations" under Patent No. 789449 dated *May 09, 1905*,

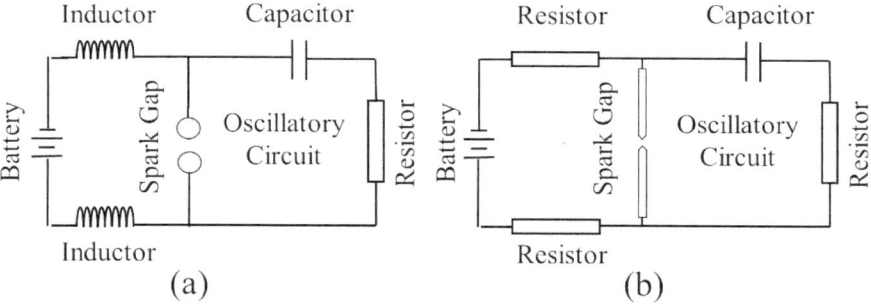

Fig. 5.1 The circuits of **a** Elihu Thomson's singing arc, **b** Duddell's singing arc

Fig. 5.2 Transmitter based on Valdemar Poulsen's work installed at Copenhagen, *Source* CC BY-SA 3.0, Ellery W. Stone |January 01,1919| elements of radiotelegraphy, D. Van Nostrand Co., New York, facing p. 165, Plate 25 on Google Books

in which arc is produced by self-induction that is subjected to atmosphere containing hydrogen at a pressure different from atmospheric pressure.

Valdemar Poulsen's invention [74] opened the doors for developments of arc-based radio telephone. By the end of *1903*, spark transmitters were successfully experimented with fairly satisfactory results in wireless telegraphy. While from the beginning of *1904*, the spark transmitters were being used for voice in addition to telegraphy and efforts to make it perfect continued till the end of *1904*. Valdemar Poulsen technology was transferred to Germany and Britain in *1906*. Cyril Frank Elwell [75] brought this technology of Valdemar Poulsen to USA in *1909*. Although Poulsen's technology was used in Europe and America, Cyril Frank Elwell found many problems in Poulsen's technology, who in later year in *1912* established Federal Telegraph Company. Based on Poulsen's work, a transmitter was built by William Duddell and succeeded in installing it in Copenhagen as shown in Fig. 5.2.

5.1.2 Starting of an Era of Continuous Wave for Wireless Telephony

A Canadian physicist Reginald Aubrey Fessenden's uncle played an important role in grooming inquiring mind of Reginald Aubrey Fessenden by taking him to witness the demonstration of Bell's homestead on *August 4 1876*. In discussion with his uncle Cortez Fessenden, he posed a question to his uncle "Uncle, How do you think the roar of thunder can be heard?" and left it for Reginald Aubrey Fessenden think and solve it.

5.1.2.1 Idea of Continuous Wave (CW)

Later, at the age of 31 years on *1897*, he threw a stone into Chemung Lake, Peterborough, Ontario, Canada and keenly observed the phenomenon of waves radiating from the point of stone's strike, posing a question to his uncle in discussions "Can't you thing of sound can exhibit similar phenomenon?" and tried to equate the similarity Hertzian waves radiating from antenna those keep going on and on in the form of steady stream to reach to encircle the antenna of receiving station that he described as "Continuous Waves". This idea has also been mentioned in his US Patent No. 706737 dated *August 12 1902* on "Wireless Telegraphy".

During the period from *1887* to *1890*, R. A. Fessenden worked at Edison's Laboratory at East Orange, New Jersey, United States. Further, due to financial difficulties, he worked in company called Westinghouse and left it with good relations. In *1892*, he accepted a chair of Electrical Engineering at Purdue University and stayed only for a brief period but his importance is still felt even today.

5.1.2.2 Fessenden's Efforts to Generate Continuous Sparks for Sustained Transmission of Sound Waves Using Interrupters

In pursuance of his ideas of "Continuous Waves", R. A. Fessenden was convinced that the continuous train of damped oscillations produced by Elihu Thomson from arc in *1892* can be extremely useful for practical wireless telephony systems. Fessenden started his work based on Elihu Thomson's patent (Patent No. 363185, *May 17, 1887*) entitled "Alternating Current Electric Motor" and came out with his own patents entitled "Electric telegraphy" (US Patent No. 706736, *12 August 1902*), "Wireless Telegraphy" (US Patent No. 706737 dated *August 12 1902*) and "Signaling by electromagnetic waves" (US Patent No. 730753, *June 09 1903*). In *1893*, University of Pittsburgh persuaded him to accept same chair because of George Westinghouse wanted him to be nearer. For this position he received substantial honourarium of $1000 that helped him be in better financial position.

Fessenden directed his efforts to provide repetitive sustained spark transmission, He started experimentation on spark transmitter [76] and in the process, with the

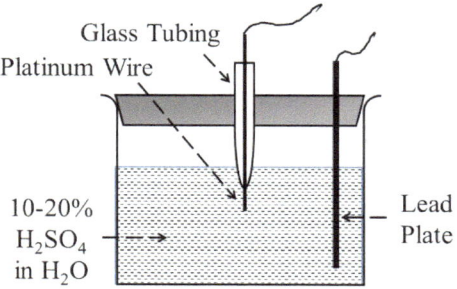

Fig. 5.3 Wehnelt's interrupter

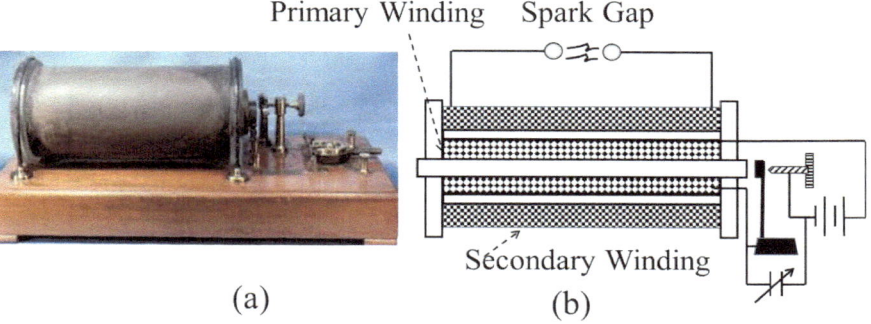

Fig. 5.4 Ruhmkorff's coil **a** Photograph of Ruhmkorff's coil, Credit: Department of Physics, Kenyon College Gambier, Ohio 43022 USA, **b** Circuit using Ruhmkorff's coil

intension to use Arthur Wehnelt interrupter [77–80] for his spark transmitter as shown in Fig. 5.3.

Fessenden started experimenting with Wehnelt interrupter operating with Heinrich Daniel Ruhmkorff's coil [81] as shown in Fig. 5.4 in *1899* and noticed that when telegraph key is held down for long, Wehnelt interrupter produced a wailing sound producing good quality speech at receiving telephone.

The interrupter with 10000 breaks per second designed by professor Kintner was delivered to Fessenden at the beginning of *1900* for his experimentation. The transmitter was modulated using voice signals picked from the microphone that was directly inserted in series with antenna as shown in Fig. 5.5.

The repetitive spark wave basically required for carrying continuous incoming voice from microphone is shown in Fig. 5.6.

At receiver side, signal picked using receiving antenna was detected using point contact screw adjustable electrolytic rectifier constructed using sulphuric acid (H_2SO_4) also called "Liquid Barretter (US Patent No. 727331, dated *April 09, 1903*)" as shown in Fig. 5.7.

Both transmitter and receiver were tried for voice transmission at the distance of 1.6 kms and with many unsuccessful attempts, finally at Cobb Island, Maryland, Vir-

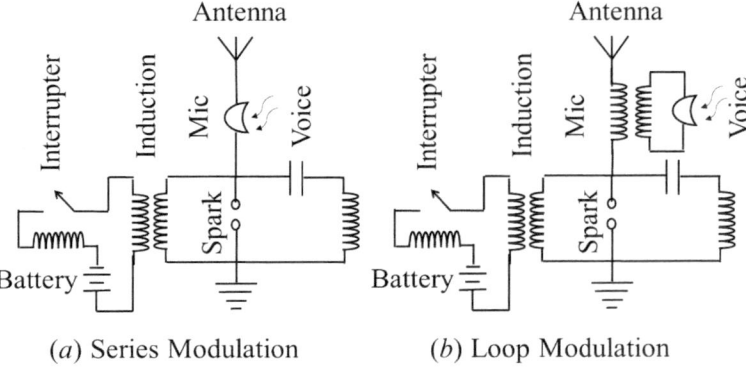

(a) Series Modulation   (b) Loop Modulation

Fig. 5.5 Reginald Aubrey Fessenden's spark transmitter: the series and loop modulation of signals from microphones

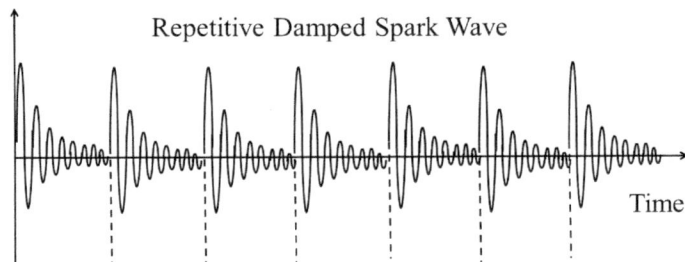

Fig. 5.6 Reginald Aubrey Fessenden's spark transmitter for voice transmission makes use of repetitive damped spark wave

Fig. 5.7 **a** Fessenden's barretter. **b** Reginald Aubrey Fessenden's receiver using electrochemical interrupter (liquid barretter)

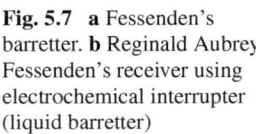

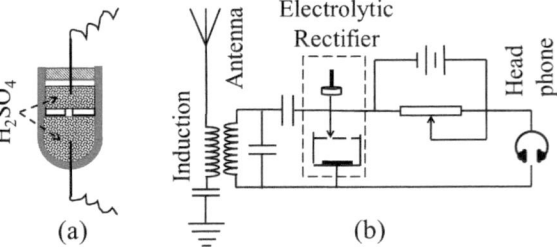

ginia, on *December 23, 1900* using antennas formed using two masts with distance of 15 m between them, the attempt was successful, whose photograph is as shown in Fig. 5.8 and this was the first time in the world that intelligible speech was transmitted using electromagnetic waves. In the conversation, Fessenden said "Hello" to other parties, was a type of successful voice conversation [82]. This invention opened a way for voice radio broadcasting.

Fig. 5.8 Reginald Aubrey
Fessenden's experiment at
Cobb Island, *Source*
Communication Canada

Fig. 5.9 Fessenden's early
version of synchronous
rotary spark gap transmitter,
Credit CC BY-SA 3.0 G. V.
Buck |2015| cyclopedia of
applied technology, Vol. 7,
American Technical Society,
Chicago, 1919, facing p. 155
on Google Books

5.1.2.3   Fessenden's Experiments to Generate Continuous Synchronous Rotary Sparks Generation for Sustained Transmission of Sound Waves Using Alternators

Although Fessenden was successful in transmitting voice continuously using interrupter, he was unhappy with the arrangements of continuous generation of discrete
bunches of sparks such that next spark is generated before earlier spark dies down
and thought of another mechanism of spark generation called "rotary spark generation" leading to development of synchronous rotary spark gap transmitter. The
early version of Fessenden's invented a spark transmitter called "Rotating spark gap
transmitter" in *1899* is shown in Fig. 5.9.

In his "Rotating spark gap transmitter", AC generator was used to provide energy
to spark gap transmitter that was directly coupled to rotary spark gap mechanism
such that spark gaps coincided with the peak of the AC cycle. The circuit diagram of
Fessenden's early version of "Synchronous rotary spark gap transmitter" is shown
in Fig. 5.10.

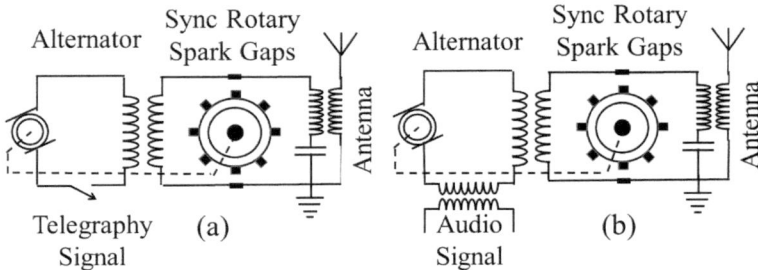

Fig. 5.10 Circuit diagrams of Fessenden's early version of "synchronous rotary spark gap transmitter": **a** circuit for transmission of telegraphy signal, **b** circuit for transmission of telephony signal

Fessenden's pioneering work [83] provided timely needed shift in favour of audio signal communication technologies from earlier Marconi's damped-wave-coherer system mainly served the interest of telegraphy. Reginald Fessenden was interested in sending alternating currents at varying frequencies, by doing so he was targeting wireless transmission audio frequency/voice through the space [84].

In *1891*, he started experimenting alternating currents and developed "An alternating-current dynamo" to continuously generate pure sine waves of uniform strength of high frequency called continuous wave radio signal transmitter, for which he applied for a patent in *May 29, 1901* that was issued in *1902* (U.S. Patent 706737).

5.1.2.4 Invention of Heterodyning Principle by Reginald Aubrey Fessenden

In *1902*, Reginald Aubrey Fessenden provided stable and sensitive receiver for radio reception called Baretter, that produced clicks when Morse key is closed or opened which was more suitable for transmission side than receiver side and was not suitable for receiving telegraphic signals, adopted by US navy for its communication needs. Many inventors directed their efforts towards the improvements of receiver sensitivity. R A Fessenden invented radio frequency signal processing techniques called "Heterodyning" that produces new frequency by mixing two more frequencies. The mix that is produced is a lower frequency called beat frequency of intermediate frequency. The envelope of beat frequency is nothing but the signal to be detected. The main advantage of heterodyning is complete encoding of all frequencies or band of frequencies contained in incoming signal. The heterodyning principle is shown in Fig. 5.11.

On *Aug 12, 1902*, R A Fessenden took "wireless signaling" US Patent No. 706740 that was a basis of heterodyne principle and is used to generate new frequencies by mixing two or more signals. In this invention, he used two or more radiating sources to generate persistent waves or impulses of different periodicities and have different characteristics Fig. 5.12a, while at receiving end Fig. 5.12b, corresponding number

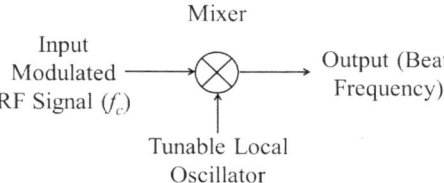

Fig. 5.11 Principle of heterodyning

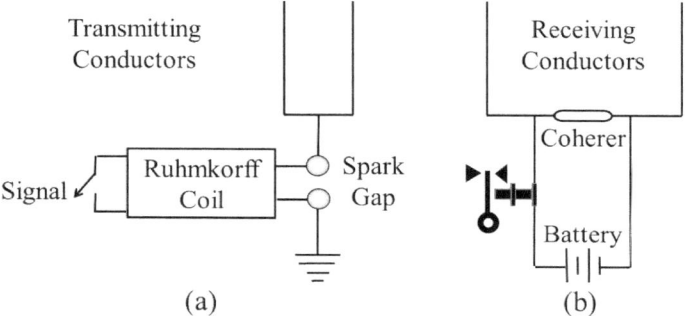

Fig. 5.12 Heterodyning circuits of Fessenden

of tuned or adjusted conductors or wires are used and their combined action produces oscillations of predetermined phase difference. These circuits can be modified for high capacity using large areas for transmitting and receiving instead of simple wires. By inventing heterodyne principle, Fessenden possibly might have targeted the processing of voice signals consisting of various frequencies in the band from 300 to 3400 Hz, having varying amplitudes and frequencies. Transmitting such signal using an antenna designed for peculiar frequency characteristics is an inefficient way of transmitting such voice signal while various frequencies should be actually catered to. Hence, Fessenden used two antennas suitable for two different frequency characteristics to transmit input signal effectively and at receiving end corresponding matched antennas, efficiently pick up respective signals which are mixed together to get transmitted signal reconstructed back.

5.1.2.5 Fessenden's Efforts to Acquire High Frequency Alternators for Transmission of Telephony Signals

In *1903*, Reginald Fessenden approached General Electric Company (GEC) for producing series of high frequency continuous wave (CW) transmitters and received 10 KHz version that he could not use for transmission of audio signals due to fact that range of audio signal is up to 20 KHz. GE supplied second alternator in *1905*, indicating their inability to provide the alternator that can be operated beyond 10 KHz.

(a) Brant Rock Radio Station   (b) Brant Rock Transmitter

Fig. 5.13 Fessenden's historic broadcast from Brant Rock station: **a** Brant Rock station, Source: CC BY-SA 3.0, Penny postcard by unknown author |1910| Reginald Fessenden's Brant Rock, Massachusetts radio tower. In 1906, Fessenden used this radio station to make the first two-way transatlantic radio transmission, communicating with an identical station in Machrihanish, Scotland, **b** Radio transmitter at Brant Rock station, Source: CC BY-SA 3.0, Jonathan Adolf Wilhelm Zenneck |1915| wireless telegraphy, McGraw-Hill Book Co., New York, p. 205, fig. 251 on Google Books

Altough Fessenden had the problems with supplied alternator, he contructively used its parts to design his own half kW machine that operated at frequencies as high as 75 KHz [85] and his problem of using high frequency continuous carrier wave was over which audio signals can be modulated. Later machine reached up to 250 kHz and 250 kW specifications. In *August 1906*, GEC delivered 50 KHz version of continuous wave (CW) transmitter. Although he could achieve the goal of transmission of audio signals, he could still face the problem of weak audio signals due to lack of availability of mechanism of amplification of weak audio signals.

5.2   Fessenden's Historic Broadcast from Bant Rock Station

The three days before the occasion of Christmas eve on *December 21, 1906*, Reginald A. Fessenden made his historic wireless transmission [86, 87] by making extensive arrangements. The US Navy ships which were previously fitted with Fessenden's radio sets were capable of receiving amplitude modulated (AM) signal, and United States Fruit Company ships notified regarding Christmas eve program broadcast from Brant Rock Station. The broadcast on *December 24, 1906* showed that human voice and music can be wirelessly transmitted using 100 KHz CW version of AC dynamo from Brant Rock, Massachusetts, USA to a Plymouth, Massachusetts, USA covering a distance of 10.34 miles. This was the experimental broadcast of voice and music on the eve of Christmas by playing "O Holy Night" on his violin. This first historic wireless telephony transmission of R. A. Fessenden on Christmas paved a way for mushroomed growth of popular public radio broadcast systems. The Fessenden's historic broadcast from Brant Rock station is shown in Fig. 5.13.

Although Fessenden was the first successful person to practically and flawlessly transmit the voice by modulating it over contiguous carrier wave, AM technique was existed in the form of experimental work done by Mayer (*1875*) [88–90], Leblanc (*1886*) [91, 92], Rayleigh (*1894*) [93] and Landell de Moura [94–97] (*October 11, 1904*).

5.3 Conclusion

This chapter stresses the events of efforts leading to the developments of wireless voice communications and the contributions made by Elihu Thomson, Valdemar Poulsen and Reginald Aubrey Fessenden. The initial efforts were directed towards continuous spark generation, continuous rotary spark generation and continuous synchronous rotary spark generation. The real twist came when Fessenden brought in the concept of continuous wave that evolved the concepts of heterodyning and amplitude modulation that drastically improved wireless voice transmission.

Chapter 6
The Era of Vacuum Tube Developments and Spread of Wireless Telephony

Abstract Fessenden's work was a milestone in telephony that generated great interest in general public, however, needed many improvements in detection and amplification of signals. This chapter describes the role of thermionic devices in the promotion of wireless telephony systems from the perspective of efficient use of continuous waves as a carrier of sound waves. The thermionic emission was basically carved out of research pursued on the production of light using electric current many personalities like Sir Humphry Davy, James Bowman Lindsay and Frederick Guthrie discovered that red hot or thermally heated conductor can easily lose negative charge but can hold positive charge. John Ambrose Fleming working under James Clerk Maxwell was influenced by Frederick Guthrie also started working in thermionic emission. The invention of the Edison effect provided a mechanism of flow of negative charge. John Ambrose Fleming used the Edison effect in radio detection and constructed a device called Fleming's Valve. Although Edison and Fleming were successful in demonstrating thermionic emission practically, Lee de Forest adopted a more systematic path to study thermionic emission and invented a device called "Audion" capable of providing ampliation to the detected signal. These devices provided great service for wireless transmission of voice and music and provided a great boost to broadcasting and become also hobbyists business. The amateurist assigned their own call signs to their radio transmitters and the registry was maintained and published by "Wireless Association of America", This also layed the foundations for the regulations of radio transmitters and the US Department of Commerce took initiative in the direction of licensing. Radio Act of August 1912 came into force with effect from December 13, 1912, for assigning call signs to radio transmitters. This act included the licensing of radio transmitting apparatus and divided radio frequency spectrum for licensing in 187.5 to 1500 KHz wavebands. The invention of regeneration, autodyne and superheterodyne principles greatly improved receiver performance and radio broadcasting widely spread after world war I and during this world war, wireless radio systems were also being used for direction finding and finding enemy targets.

6.1   Rise of Amateur Radio: Primitive Noncommercial Unorganised Experimental Wireless Audio Broadcasting

The people started pursuing wireless telegraphy study as a hobby or sport and were responsible for experimenting, developing and spreading amateur radio sets based on Hertzian waves. Although it is difficult to tell that who was the fist amateur radio operator?, British citizen Leslie Miller is considered to be a sure contender [98] for this honour and he was the first person to publish his article entitled "simple-to-build transmitter and receiver" in the issue of "The Model Engineer and Amateur Electrician", published from London in _March 1898_.

It was difficult for amateurs to get the information on Hertzian wave radios and hence many magazines came in the market to support amateurist and their history traces back to _1901_ when "Amateur Work" [99] published a material regarding how to build simple wireless radio systems and progress on amateurs. In _1904_, the issue of "Amateur Work" carried the information on sets that covered the range of eight miles. In _1906_, "Technical World Magazine" published an article written by M.W. Hall mentioned about the work carried two Rhode Island teenagers. As time passed, wireless radio technology became more and more matured and in _1904_, a company called "Electro Importing Company of New York City" set up by 18 year old Luxembourg immigrant Hugo Gernsback was first to offer affordable wireless radio equipment kits for experimenters.

In _April 1909_, Charles Herrold broadcasts from his Herrold School Electronics Institute in San Jose downtown called "San Jose Calling" and then many "call signs" started appearing and hence Department of Commerce started regulating radio and Herrold was the first person to coin the terms like "Broadcasting" and "Narrowcasting".

Due to non-existent control on amateurs, Amateur radio (called Ham radio in united states) transmitters grown indiscriminately as per the will and pleasure of amateurs. During the Act of licensing of _1912_ imposed by the United States Commerce Department, amateurs were restricted to 200 m (1500 kHz) wave frequency. At that time, it was a belief that longer wavelengths give better distance performance and shorter wavelength like 200 m useless to achieve distance performance. Many started fearing that overcrowding around this useless wavelength of 200 m will eventually be frustrating and left to pursue their interest in amateur radio. However, a notable person Hiram Percy Maxim showed that messages could be transmitted over longer distances if relay stations are properly organised and cooperation amongst them is increased. Many amateurs extended their physical coverage by relaying their messages through other amateurs. In following years the number of amateurs increased and to have greater cooperation amongst them, Hiram Percy Maxim founded American Radio Relay League (ARRL) in _1914_.

After _April 7, 1917_, when US got involved in World War I, all Amateur radio transmitters in the United States were shut down for security reasons by the order of the Chief Radio Inspector of the United Sates Navy, however, expertise of around

4000 amateurs did not go unnoticed and their services were used in war as radio operators. The war ended on *November 11, 1918*, however, ban remained in force up to *October 01, 1919* when licences to amateurs were resumed and allowed to operate.

From *1919*, amateurs adopted vacuum tube technology and started exploring shortwaves which otherwise supposed to useless at that time. This resulted in an increased range and reliability of radio equipment that led to widespread developments in broadcasting.

6.2   Need for Detecting and Amplifying Devices and the Role of Thermionic Emission

Initially, it was an intent to produce electric light and Sir Humphry Davy [100–102] was the first person to state that electricity can be produced by chemical action and conversely electricity can decompose the chemical compounds into its fundamental elements, it is because chemical forces are basically electrical in nature. Sir Humphry Davy invented a carbon arc that produced light using electricity or carbon is raised to incandescence. His efforts were directed to find suitable elements that will produce electric light.

Many researchers [103, 104] like James Bowman Lindsay (*1835*) who was the first person to demonstrate his incandescent electric bulb in public meeting at Dundee, Scotland; A British scientist Warren de la Rue (*1840*) made incandescent vacuum bulb using platinum filament; Frederic De Moleyns (*1841*) was granted the first patent for incandescent bulb that used platinum filament similar to similar to Warren de la Rue; Joseph Wilson Swan (*1850*) used carbonised paper filament that had problem of blackening; Woodward and Evans patented an electric light in Canada (*July 24, 1874*); Alexander Lodygin (*1874*) was first person in Russia to hold patent for incandescent electric bulb used molybdenum filament; William Sawyer's (*1877*) lighting apparatus; Joseph Swan patented incandescent bulb *1878*; Lewis Howard Latimer (*1882*) used carbon filament and his patent was purchased by Edison and was also hired by Edison in *1884*.

Invention of thermionic emission [105], which is a basis of developments of detecting and amplifying devices, was carved out of research pursued on the production of light using electric current or electric bulbs. Thermionic emission is a process in which charged particles are emitted from the surface of the conducting body when external heat energy is supplied. When external heat energy is applied to the conduction bodies, charged particles are emitted. The more the heat is supplied more the charged particles are emitted from the surface of the conducting and these charged element particles are taken off from the surface when a suitable electric field is applied.

The charged particles may be electrons or ions depending on the type of emitting surface. When heat is supplied to the conducting bodies like metals, valence electrons of the atoms of the metal, gain the energy and break free from the atoms of the metal by breaking its bonding with parent atoms of the metal and hence called as free electrons, more the heat is supplied, more the free electrons. The minimum temperature at which the process of emission of free electrons starts is called threshold temperature.

Thermionic emission was first reported by Frederick Guthrie _1873_ [106] and he discovered that red hot (thermally heated) conductor (iron sphere) can easily loose negative charge but can hold positive charge. At the age of 24 in _1873_, John Ambrose Fleming was influenced by the studies of Frederick Guthrie about "how hot bodies lose charges?" [107], at that time was also shown that hotter bodies lose negative charge or"charge of electricity". At the age of 28 (_1877_), John Ambrose Fleming was studying in Cambridge and conducted his research under James Clerk Maxwell. In _1882_, he accepted an assignment of consultant in Edison Electric Light Company of London. In _1884_, he became a chair of electrical technology in University College, London and worked there for 40 years, from _1899_ he also worked as scientific advisor to Marconi Company.

Thomas Alva Edison started on working on incandescent element using carbonised cotton thread or coiled thread or sheets as his filament was clamped between platinum wires and placed this mechanism in an evacuated glass bulb that was tested on _October 19, 1879_ and worked successfully for two days and on _October 21, 1879_ it burnt out, this lamp was considered to be first practically feasible commercial lamp and was revised many times. For his invention of electric lamp, he got US Patent No. 223898 on _January 27, 1880_ as shown in Fig. 6.1.

While experimenting with incandescent light bulb, carbon soot given off by carbon filament got deposited in the inner surface of the bulb caused darkening of the bulb, and in order to solve this problem, Edison inserted an additional electrode in the bulb space that was connected to the positive terminal of the battery, thinking that the electrode will attract the soot emitted by carbon filament. To his surprise, the

Fig. 6.1   a Thomas Alva Edison's US Patent 223898 Dated January 27, 1880 of Electric Lamp and **b** Photograph of Edison's Electric Lamp   *Credit* The Thomas A. Edison Papers at Rutgers University

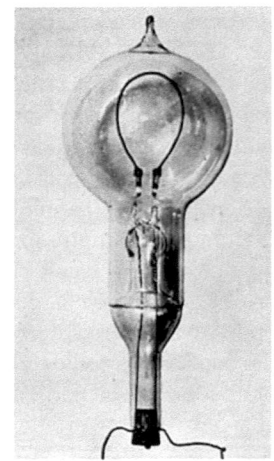

(a) Edison's Patent 223,898 Electric Lamp

(b) Photograph of Edison's Electric Lamp

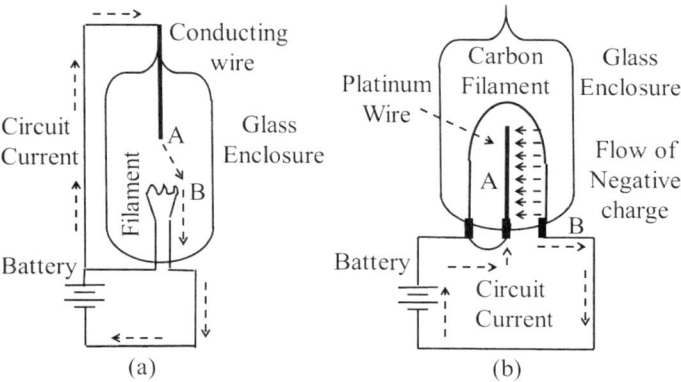

Fig. 6.2 Apparatus for studying Edison Effect

arrangement that he provided, served as an additional current circuit that provided a current that was dependent on the hotness of filament. This observation gave birth to the "Edison Effect".

6.2.1 Edison Effect

The classical example of thermionic emission is the emission of electrons from hot cathode in to a vacuum (also called as Edison effect) [108], enclosed within the glass cover called vacuum tube discovered by Thomas Edison on *February 13, 1880*. The Edison applied patent for his "Edison effect" in *November 15, 1883*, and US patent No. 307031 was granted in *October 21, 1884*. The Edison effect can be explained using Fig. 6.2.

The Edison Effect states that *"when a conductive substance is interposed in a space within the globe of incandescent lamp that is connected to a positive terminal of the battery and when the lamp is in operation, then current flows in shunt circuit formed by the A point of wire and negative terminal B of incandescent element".* That means, electrons from filament were attracted to point A to establish a flow of electrons to a positive terminal of the battery, but the term electron was not used at that time and was introduced by J.J. Thomson in *1899*.

In *1885*, W. H. Preece presented his work before Royal Society that was similar to that of Edison's work entitled "On the Peculiar Behaviour of Glow Lamps when Raised to High Incandescence".

Fig. 6.3 Fleming Diode or
Fleming Valve—Converts
alternating currents into
continuous currents

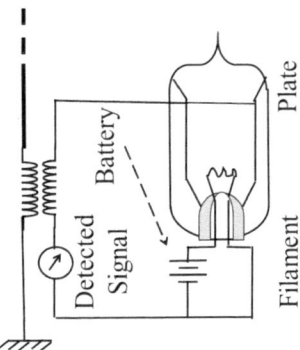

6.2.2 Fleming's Valve or Fleming's Diode Valve

In _November 16, 1904_ John Ambrose Fleming a former employee of Edison and
current employee of Marconi Company at that time found that the Edison effect can be
used in a radio detection. Fleming, FRS and a Pender Professor at UCL [109] patented
his thermionic device in Britain as British patent No. 24850 in _November 1904_ and
further in _November 7, 1905_ [110] he patented a device called "Fleming Valve" in
US as US Patent No. 803684 for radio detection shown in Fig. 6.3.

Although Edison and Fleming were successful in demonstrating thermionic emission practically, later on, Lee de Forest adopted a more systematic approach to study
the theory behind thermionic emission and invented a device called "Audion" that
was successfully demonstrated for the detection of Hertzian waves.

6.2.3 Invention of Audion

Lee de Forest's entry in wireless radio communication put this discipline on a dis-
tinctive path when he started experimenting with electrolytic responder in _1900_, that
put him on a track of good fortune. He experimented with a solitary gas burner
with Welsbach mantle [111] and as usual, he generated a spark using induction as
a source of Hertzian waves. When the spark generated, his flame-based detector as
shown in Fig. 6.4 responded. Lee de Forest observed that the light from the incandescent mantle (M) diminishes whenever coil sparks and by proper adjustment of gas
and air, complete extinction of light from mantle was observed during the sparking
of the coil. This effect increased the curiosity of Lee de Forest to investigate this
phenomenon while other adjustments increased the light above the normal. He was
tremendously delighted over this invention of a sensitive Hertzian waves detector.
In this experiment possibly, he might also have observed the fluctuations in flame
created due to sparking of the coil which was clarified in his later descriptions of his
experiment that when he enclosed his sparking mechanism in a wooden closet the

Fig. 6.4 Flame based
detector for detector for
Hertzian waves

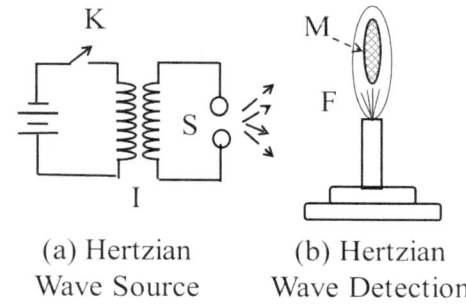

(a) Hertzian
Wave Source

(b) Hertzian
Wave Detection

Fig. 6.5 Flame based
electric circuit for the
detector of Hertzian waves

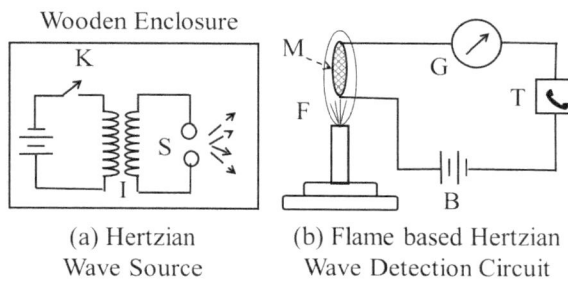

(a) Hertzian
Wave Source

(b) Flame based Hertzian
Wave Detection Circuit

flame ceased to fluctuate and was illusioned by the observation that light from the spark was obstructed due to the enclosure around the spark mechanism and he could not realise that flame which in most susceptible to movements and in a tenuous state, can be extremely sensitive to sound and heat.

However he was not really convinced about his experiment and to verify that whether it was really due to blocking of light or electromagnetic waves or sound created due to spark?, he soaked his mantle (M) in potassium or sodium solution and dried to place it in a circuit containing Battery (B), Galvanometers (G) and Telephone receiver (T) as shown in Fig. 6.5. In this experiment, he got the faint response to Hertzian waves, and with the experiment, Lee Forest was now convinced that it was visible light and sound that were blocked and not the electromagnetic or Hertzian waves.

With this success, motivated Lee Forest started inventing different ways to electrify the luminous portion of flame to make it sufficiently conducting using salts of alkaline metals like sodium, potassium and caesium. Arrangement was made to hold platinum cup containing alkaline metal salt solution in a flame and luminous part of flame was made as cathode (K) of telephone circuit and anode (A) was made up of platinum wire or a plate just held above the cup. This arrangement was an early form of "Audion" was sensitive to feeble high frequency and supplied a current in the order of several milliampere through the circuit using coloured flame, that produced exact sound of spark in telephone. The salt was electrolyzed. It was observed that the current does not follow the electromotive force, that is relation $(V = IR)$ or Ohm's

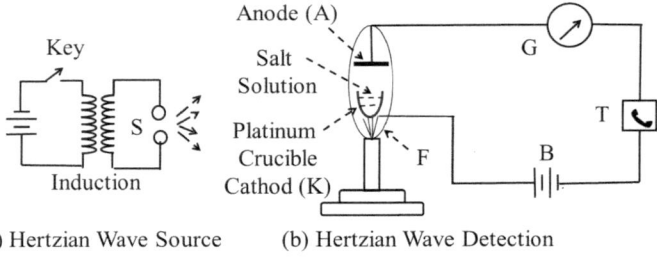

Key

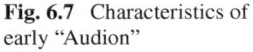

Anode (A)

Salt
Solution

Platinum
Crucible
Cathod (K)

Induction

(a) Hertzian Wave Source

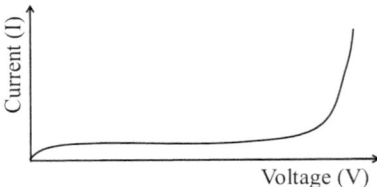

G

T

F   B

(b) Hertzian Wave Detection

Fig. 6.6 Flame Electrification using a concept of anode and cathode

Fig. 6.7 Characteristics of
early "Audion"

law does not hold good and its maximum value is limited to its saturation value. This
early form of "Audion" is shown in Fig. 6.6.

The amount current that can be passed through the aqueous solution is equivalent
to the amount of salt electrolysed and converted into heated gas in the form of salt
vapour. The operational characteristics of this early form of "Audion" is shown in
Fig. 6.7.

The salt vapour at atmospherical pressure containing positive heavy corpus-
cles(ions or tiny particles) and negative corpuscles, the negative corpuscles being
light, the velocity of negative corpuscles increases rapidly with temperature and electric potential gradient (volts/centimeter). Lee Forest computed this parity of velocities of positive and negative corpuscles and temperatures at 1000 °C and 2000 °C and
found that velocities of negative corpuscles are 3.715 times and 40 times respectively
as that of positive corpuscles.

The velocity of negative corpuscles is proportional to electric force applied
between cathode (K) and anode (A) as shown in Fig. 6.6. The distance between anode
and cathode (incandescent body) is set to be half wavelength of electric oscillation
as used in wireless telegraphy.

If Hertzian waves or oscillations as passed over the flow of hot gas consisting
between cathode and anode, impose its momentary potentials on moving ions and
interfere their motions changing the currents through the circuit are sensed by telephone receiver, such changes can even be sensed even if the battery is not used in
the circuit.

It was then time to replace gas flame by some electrical mechanism that will
behave like flame environment or to produce incandescence that is characterised

by the production of light by raising the temperature of an element for sensing the hertzian waves.

In 1882, Julius Elster and Hans Friedrich Geitel [112] developed a "Theory of electric processes in thunderstorms" wherein, they made a systematic study of ionisation and published *Influenztheorie*, that lead to work on the production of incandescence by use of electric current. They published a series of papers without any background knowledge of electrons and ions, The second paper (*1883*) [113] generation of electricity using gases in contact with incandescent elements, the third paper (*1885*) [114] in series focuses on unipolar conduction property of heated gases as light ions have high velocities and the electric current is due to negative ions. The fourth paper (*1887*) [115] is especially dedicated to electrification aspects of gases due to use of incandescent elements, while last paper (*1899*) in series [116] highlights electricity generation by rarified gases in contact with electrically heated wires. Elster and Geitel's work was a motivation to researchers like Sir Joseph John Thomson.

In *1899*, Sir Joseph John Thomson made a study of ions in gases at low pressures [117, 118] and found that in a low pressure, the carbon filament emits fundamental light negative particles, which are the part of atoms, and with this study, Sir Joseph John Thomson [119, 120] introduced a concept of electron and showed that negative ions are emitted from carbon filament placed in low pressure (fairly good vacuum) environment are nothing but electrons.

The electrons moving through the gas enclosed in a glass tube [121] are also called as cathode rays and such tube is called as cathode ray tube. With this invention electron, electron theory of metallic conduction was progressed considerably by Thomson [122], Riecke [123] and Drud [124] to conclude that "conductivity in metallic elements is due presence of an electron atmosphere in them.

In *1907*, Lee de Forest [125, 126] patented two electrode thermionic electronic device first of its own kind called "Audion". The device was able to detect and also amplify radio frequency oscillations. The functionality of "Audion" characterised by amplification of feeble radio frequency oscillations was characterised as "Vacuum Tube Triode Device [127]". Lee de Forest patented this thermionic device under the tile "Space Telegraphy" and was issued a US Patent No. 879532, Dated *Feb 18, 1908* that was filed on *Jan 29, 1907*. The structural and functional aspects of "Audion" are shown in Fig. 6.8.

The thermionic vacuum tube device in the circuit as shown in Fig. 6.8b has an ability to amplify the feeble signal and became popular as "grid Audion" and became three electrode thermionic device called vacuum tube triode. The photograph of grid audion is shown in Fig. 6.9.

The thermionic vacuum tube device of Fig. 6.8a simply acted like detection device that separated audio signal from radio frequency acted like rectifier and had no ability to amplify the feeble signals and hence became popular as two electrode thermionic device called as vacuum tube diode.

These devices rendered great service for wireless transmission of voice and music signals in audio range and became popular devices in radio broadcast systems for improving the signal quality.

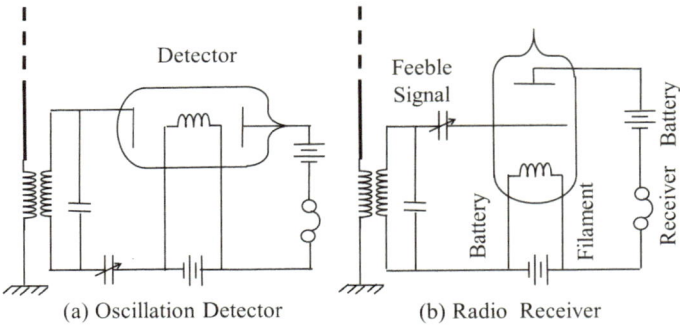

Fig. 6.8 Two types of uses of two electrode thermionic device called "Audion" **a** RF detector and **b** Space or wireless telegraph receiving system

Fig. 6.9 The photograph of grid audion *Credit* Department of Physics, Kenyon College Gambier, Ohio 43022 USA

Besides the invention of "Audion", Lee de Forest was also interested in the broadcasting business and in *1908*, he broadcasted music from the Eiffel tower of Paris for the demonstration of arc transmitter.

6.2.4 Amateurist's Self Assigned Call Signs to Their Radio Transmitters as Registered by Wireless Association of America

In *1909*, a magazine called "Modern-Electrics" published a list of radio stations with self assigned call signs with letters varying from 3 to 1 alphanumeric letters in which numerals are less. The registry was maintained and published by "Wireless Association of America" in the form of official blue book. The sample of the names registered for wireless call signs is shown in Table 6.1.

Table 6.1 Self assigned call signs registry in blue book by Wireless Association of America

Name	Call sign	Wave length (Meters)
G.C. Carpenter, New York	G.C.M.	300
Harry Atkinson, Philadelphia, Pa.	H.3.A.	65
D.K.Caldwell, Hollywood, Cal.	P.25.	200
Fred. Small, Baker City, Ore.	C.T.	100
R.A. Egbert, Alpine, Tex	R.A.	50
Fred. Frerichs, Monogahela, Pa.	L.D.M.	400

6.2.5 Laying of Foundations for Regulation for Radio Transmitters

The early radio communication served basic communication needs like health, commerce and well being of the people. Initially, communication was extensively used in urgent or emergent messages, ship navigation guidance, communication between land stations and ships, ship to ship communications and so on. The wireless communication was extremely useful in distress and SOS (Save our souls) signal was first adopted by German Government in *April 1, 1905* and become worldwide standard in *November 3, 1906* adopted internationally in maritime radio from *July 1, 1908* to *1999* which was then used as vidual SOS signal by Global Maritime Distress and Safety System standard. In *1908*, SOS (Save our souls) signal was internationally accepted as a distress call for ships using wireless systems of Marconi.

In *January 23, 1909*, RMS Republic of White Star Liner collided with Italian Lloyds S.S. Florida near Nantucket Island [128] started sinking, Jack Binns on S.S. Republic sent a Customary Quick Despatch (CQD) wireless message for assistance and in the process, all 461 passengers and 444 crew members were saved while 6 crew members, three from each of the ship lost their life. After this episode, in *1910*, Wireless Act of *1910* came in force that All ships carrying 50 passages or more travelling over 200 mi must have wireless communication facility. Due to availability of communication facilities for ships, for safety of the people traveling on ship, Wireless Ship Act of *1910* came in force [129] which mandated that ships arriving in the United States having a capacity of more than 60 passengers must have communication equipment. To have better safety, unplanned growth had to be tacked and controlled, and hence the US Department of Commerce took initiative in this direction.

However, before US Department of Commerce took initiative in the direction of licensing, foundations were laid as early as in *1906*, Berlin International Wireless Convention tried to focus on the issues of identification of transmitting radio stations using three letters, could not succeed due to ill cooperation and individual wireless operators continued to use their own method of identification. The Berlin convection legalised a term "Radio" that describes the "Wireless" better.

This was the period when there was no control over radio operators and in absence of licensing, there was no concept of call signs and issuing of call signs started only

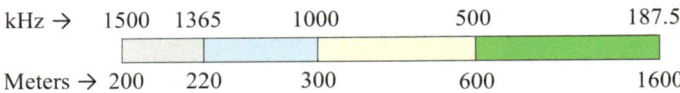

kHz →   1500   1365   1000   500   187.5

Meters →   200   220   300   600   1600

Fig. 6.10 Wave Bands as per Radio Act of August 13, 1912

after when Act of *1912* was established. Prior to the Radio, pioneers made their calls using their initials. For example, Hiram Percy Maxim a founder of American Radio Relay League (ARRL) used SNY as a call sign in *1911*.

On the night of April 14 to the morning of *April 15, 1912*, RMS Titanic on its way from Southampton to New York City strikes a huge iceberg in the Atlantic ocean, however, it's nearby ship could not attend to SOS call due to the fact that its radio operator was off the duty and the worst tragedy could not be avoided. The sinking of Titanic and the tragic loss of many lives, urgently demanded a need for a comprehensive Radio Act that could provide safety to the ships.

International Radiotelegraph Conference in London on *July 5, 1912* persuaded the formalisation of three letter call signs as proposed in Berlin. International Wireless Convention [130] formalised the beginning letters K and W for US ships and other letters for the vessels from other countries. Based on these conventions, the Radio Act of *August 13, 1912* [131] came in force with effect from *December 13, 1912*. This act gave the powers to the Bureau of navigation in the US department of commerce, to assign call signs for US ships and offshore stations and their assignment started with letters K(for West) and W(for East). The act also included the licensing of radio transmitting apparatus and could not mention radio broadcast. It formulated the regulations for radio apparatus (includes apparatus for ship stations and land stations)and radio operators (includes commercial, Armature and Technical). It divided the radio frequency spectrum for licensing purpose as shown in Fig. 6.10.

From the last or final band that is from 1365 kHz (220m) to 1500 kHz (200 m) really a single frequency band assigned to amateur stations and was considered to be useless and hence was removed due to its poor ground coverage that was used by amateurs prior to *1912*. Further, it was opined that instead completely scrapping this band, it can be utilised by naval applicatiions, while special licences to be issued to the amateurs for 200 m (1500 kHz) to 600 m (500 kHz) waves.

Department of Commerce, United States, started issuing call signs only after *1912*, for various users in all 9 radio districts and all radio operators required licences for sending the signals on the air for communication or broadcasting entertainment programmes.

Other stations were classified under "Amateur and Special Land Stations" covering nine radio districts. The call sign was formulated by radio district number (from 1 to 9) plus two alphabetic characters. These radio districts were as shown in Table 6.2.

As per Act of *August 13, 1912*, call letter policies were formulated by US Department of Commerce on *May 9, 1913*, and circular regarding call letters for various countries was issued on *April 23, 1913* as shown in the Table 6.3.

Table 6.2 United States, Radio Districts for Amateur and Special Land Stations in US

Radio District Number	Areas included
Radio District No. 1	Boston, Ma., Maine, New Hampshire, Vermont, Massachusetts, Rhode Island, Connecticut
Radio District No. 2	New York, N.Y... (county of New York, Staten Island, Long Island, and counties on the Hudson River to and including Albany, Rensselaer, and Schenectady) and New Jersey (counties of Bergen, Passaic, Essex, Union, Middlesex, Monmouth, Hudson, and Ocean)
Radio District No. 3	Baltimore, MD...New Jersey (all counties not included in the second district), Pennsylvania (counties of Philadelphia, Delaware, all counties south of the Blue Mountains, and Franklin County), Delaware, Maryland, Virginia, District of Columbia
Radio District No. 4	Savannah, GA...North Carolina, South Carolina, Georgia, Florida, Porto Rico
Radio District No. 5	New Orleans, LA...Alabama, Mississippi, Louisiana, Texas, Tennessee, Arkansas, Oklahoma, New Mexico
Radio District No. 6	San Francisco, CAL...California, Hawaii, Nevada, Utah, Arizona.
Radio District No. 7	Seattle, WASH...Oregon, Washington, Alaska, Idaho, Montana, Wyoming
Radio District No. 8	Cleveland, OHIO...New York (all counties not included in the second district), Pennsylvania (all counties not included in the third district), West Virginia, Ohio, Michigan (Lower Peninsula)
Radio District No.9	Chicago, ILL...Indiana, Illinois, Wisconsin, Michigan (Upper Peninsula), Minnesota, Kentucky, Missouri, Kansas, Colorado, Iowa, Nebraska, South Dakota, North Dakota

Irving Vermilya was considered to be the first licensed radio operator with 1ZE as a call sign and was got Skill Certificate No.1 in *1912*.

The annual list of stations in US were published Radio Service Bulletin from *1913* to *1916* which was suspended during world war 1 (*28 Jul 1914–11 Nov 1918*) and restarted in the beginning of *1919*.

6.2.6 Wireless Systems After Invention of Vacuum Tubes

In the year *1907*, "The Wireless Speciality and Apparatus Company", [132] was established by Phillip Farnsworth, G. W. Pickard, and Col. John Firth to gain access to fine wireless techniques for their ship and shore wireless stations which were also in possession of patent rights of Crystal detector for wireless applications. In *January 12, 1908*, Lee de Forest made first musical radio broadcast along with his wife Nora from Eiffel tower, and soon after this program, a permanent underground wireless transmission centre was constructed near Eiffel tower's south leg and from *May 23, 1910*, it became an official broadcasting station with appropriate call sign.

Table 6.3 Call sign assignments of various countries

Call Signs	Respective Countries
Axx, Bxx, Dxx, Fxx, Gxx, Ixx, Jxx, Mxx, Nxx, Qxx, Rxx, Wxx	Germany, Britain, Germany, France, Britain, Italy, Japan, Britain, US, Reserved, Russia, US
(CAA-CMZ), (CNA-CNZ), (COA-CPZ), (CQA-CQZ), (CRA-CTZ), (CUA-CUZ), (CVA-CVZ), (CWA-CWZ), (CXA-CZZ)	Not Assigned, Morocco, Chile, Monaco, Portugal, Not Assigned, Romania, Uruguay, Not Assigned
(EAA - EGZ), (EHA - EZZ)	Spain, Not Assigned
(HAA - HFZ), (HGA - HHZ), (HIA - HZZ)	Austria-Hungary-Bosnia-Herzegovina, Siam, Not Assigned
(KAA - KCZ), (KDA - KZZ)	Germany, US
(LAA - LHZ), (LIA - LRZ), (LSA - LWZ), (LXA - LZZ)	Norway, Argentina, Not Assigned, Bulgaria
(OAA - OFZ), (OGA - OMZ), (ONA - OTZ), (OUA - OZZ)	Not Assigned, Austria-Hungary-Bosnia-Herzegovina, Belgium, Denmark
(PAA - PIZ), (PJA - PJM), (PJN - PJZ), (PKA - PMZ), (PNA - PZZ)	Netherland, Dutch Curacao, Dutch Surinam, Dutch East Indies, Not Assigned
(SAA-SMZ), (SNA-STZ), (SUA-SUZ), (SVA-SZZ)	Sweden, Brazil, Egypt, Greece
(TAA-TMZ), (TNA-TZZ)	Turkey, Not Assigned
(UAA-UMZ), (UNA-UZZ)	France, Austria-Hungary-Bosnia-Herzegovina
(VAA-VGZ), (VHA-VKZ), (VLA-VMZ), (VNA-VNZ), (VOA-VOZ), (VPA-VSZ), (VTA-VWZ), (VXA-VZZ)	Canada, Australia, New Zealand, South Africa, Newfoundland, British colonies, India, Not Assigned
(XAA-XCZ), (XDA-XZZ)	Mexico, Not Assigned
(YAA-YZZ)	Not Assigned
(ZAA-ZZZ)	Not Assigned

Early wireless broadcast took place in the same year on *January 13, 1910*, the inventor of vacuum tube triode Lee de Forest aired the program programmes featuring the voices of Enrico Caruso and other opera stars from New York's Metropolitan Opera House. In *July 1911* Dr. Lee de Forest was chosen to head the Federal Telegraph Research Laboratory.

In *1912*, "The Wireless Specialty Apparatus Company" was purchased by "United Fruit Company" to get access to fine wireless gadgets. In the same year Australian-born Cyrill F. Elwell, of California based Federal Telegraph Company, demonstrated improved rotary spark mechanism to US Navy catering to both telegraphic messages and simultaneously can be used for voice communication, that made the headway for US Navy to use continuous waves (CW), with these happenings, Federal Telegraph Company came in limelight and suddenly became a company of national importance.

In *1913*, the US navy not only become habituated to use wireless systems, but were totally dependent on wireless systems.

6.2.7 Efforts Towards the Improvements of Valve

In *1909*, marketing of audion for public use started, this opened the doors for its use in radio broadcasting system. In *1910* AT&T identified the importance an amplifier in telephone circuits and work carried out by Harold Arnold in subsidiary company Western Electric of AT&T during *1911–1912* and developed mercury vapour discharge tube to control ionisation current using magnetic control but this tube was never produced. In *1912* Fritz Lowenstein demonstrated an amplifier sealed inside a lead box to prevent photographs to be taken, he also did not disclose any details and its performance was not also satisfactory and there was no way for Bell company to judge his invention. Later on in *October 1912*, John Stone Stone demonstrated "Audion" and also arranged demonstration by de Forest where full details were disclosed. Within a short notice, Arnold van der Bijl of AT&T turned this crude primitive device in to reliable amplifier. High vacuum in the device is achieved using Gaede molecular pump. A reliable filament was fabricated by coating the platinum strip using barium nitrate for improving emission at lower temperature and stronger electrode support was provided. In *1912*, Fritz Lowenstein took an important patent [133] of negative bias which as sold to AT&T at a price of $1,50,000. In *1913* A&T finally took the rights of Le de Forest's patent at the cost of $3,90,000 in parts including all right of telephony, telegraphy. The Valves VT-1 (General purpose triode) and VT-2 (5-watt transmitting valve was made available in *1917*) were mass produced by AT&T. During World War I development of robust and reliable valves for military applications started as per the military needs.

In *1912*, the General Electric was also interested in valve suitable to provide speech modulation for their high frequency alternators promoted by Reginald Fessenden, a brilliant engineer and pioneer of radio telephony at NESCO (National Electric Signalling Company. Ernst Alexanderson a pioneer of high frequency alternators at General Electric was in knowledge of audion developments and thought that it can be adopted for overcoming his modulation problem and General Electric which was already in the business of electric lamp manufacture, was the most suitable place for development robust audion for his applications. William. D. Coolidge [134] and Irving Langmuir had perfected manufacturing of reliable ductile tungsten filaments that gives more light output per watt of power. The task to development of valve was assigned to Irving Langmuir [135] and removed the limitations of poor vacuum and also shown that electron emission increases when the temperature of the filament is raised this results in raised space charge between cathode and anode. The increased space charge, repels electrons back to cathode and proved that anode current was less than it was predicted by Richardson's law which was in conformity of similar result was obtained by C.D. Child [136] in *1911*. This work of Irving Langmuir was published in Physical Review in *1913* which was further expanded in *1915* [137] in his popular paper "The Pure Electron Discharge".

Marconi's Wireless Telegraph Company (MWT) was not in the business of valve manufacturing and up to *1919*, it used valves manufactured by Edison Swan Electric Company (Ediswan) and Osram-Roberson Lamp Works of the British General Elec-

Fig. 6.11 Marconi's Short Distance Wireless Transmitter and Receiver, Credit: Marconi's Wireless Telephone, The Wireless Age, April, 1915, from the articles and extracts about early radio and related technologies, concentrating on the United States in the period from 1897 to 1927, United States Early Radio History

tric Company (GEC). In *October 20, 1919*, Marconi's Wireless Telegraph Company set up a joint venture called Marconi-Osram Valve Co. Ltd. to manufacture the valves and in *1920*, its name was changed to M.O. Valve Co. Ltd. (MOV) but production was done at Osram-Robertson Lamp Works.

Although Marconi's company was not manufacturing the valves, Henry J Round of the company was pursuing the company's own valve design since *1911*, but was suspended till *November 1912* due to many other commitments of the company. The design being pursued by Marconi Company was based on soft valve design of Lieben, Reisz and Strauss and hence agreement was reached between Marconi and Telefunken companies. In *1913*, the valves called Type-C (Receiver Valve) and Type-T (Transmitter Valve) were successfully produced. In *1914*, Type-TN was introduced for use in Short Distance Wireless Telephone Transmitter and Receiver (Fig. 6.11) along with Type-C valve. The other types of soft valves made by Marconi Company were Type-CA Type-CT, Type-D, Type-N, Type-LT and Type-TN. During World War I, direction finding equipment was fabricated using single Type-C valve, while Type-T valve was used in 120 W ground radio set used by Royal Flying Corp using 600–1000 m waves.

General Gustave Ferrié of French Military Telegraphic Service during World War I was responsible to develop differently constructed valve known as 'S' and Metal based on audion sample brought in France by Paul Pichon in *August 1914*, that was patented in *October 1915* and was very successful in World War-1 and had four pin base.

In *1916*, French valve produced in Britain by the name R-Valve and it was initially manufactured by BT-H and Ediswan during World War-1 and latter on by GEC,

Osram-Robertson Lamp Works, Cryselco, Cossor, Stearn and Moorhead Laboratories of San Francisco. The French Valve TM and R valves had the drawback of high internal capacitance between grid and plate which limits its use for RF amplifier for frequencies below 600 KHz. This limitation was removed by Type-Q valve designed by Capt. Henry Round of the Marconi Company in *1916*. This valve was made by Edison Swan for the Marconi Company which was later on production was transferred to MOV in *1919*. Further Henry Round designed V.24 in *1917* which was initially made by Ediswan in *1918* and transferred to MOV for manufacture in *1919*.

6.3   Invention of Regeneration Principle

In *1912*, Lee de Forest accidently connected output of audion to its input produced loud howling or hissing sound and instead of investigating it, he just turned blind eye to it. Later on it was identified as regeneration and was a base of regeneration circuit discovered by Edwin Howard Armstrong. Armstrong realised that audion circuit based on regeneration can serve as a powerful amplifier of incoming radio waves or Hertzian waves.

In the year *1912* Edwin Howard Armstrong started experimenting with audion [138] and discovered regenerative radio receiver circuit [139, 140] and demonstrated it to his friend Thomas Styles. For the want of $150 which his father would not give him due to the condition that he must graduate first, but on the advise of his uncle, he notarised his invention on *January 13, 1913* that served as a recorded date for his invention. In early *1913*, he also demonstrated his discovery for the public at the Columbia University where representatives of Marconi Company were present.

6.3.1   Patent Controversy Between Edwin Howard Armstrong and Lee de Forest

Edwin Howard Armstrong just after his graduation from Columbia University, applied for a patent on *29 October 1913*. However, on the suggestion of his patent attorney William Davis, he was made to add all important features of his invention "Audion as an Oscillator" for which he filed a separate application on *December 1913*, patent examiner argued that his invention is not different than his amplifier patent that virtually forced him to withdraw the patent application for the amplifier that weakened his case in later litigation. In November *1913*, he presented a paper entitled "The audion, detector and amplifier" that did not mention the role of feedback which exposed the fact that he could not understand the exact functioning of his circuit.

Edwin Howard Armstrong devised an "Improved Audion Receiver Circuit" on *January 14, 1914* as shown in Fig. 6.12.

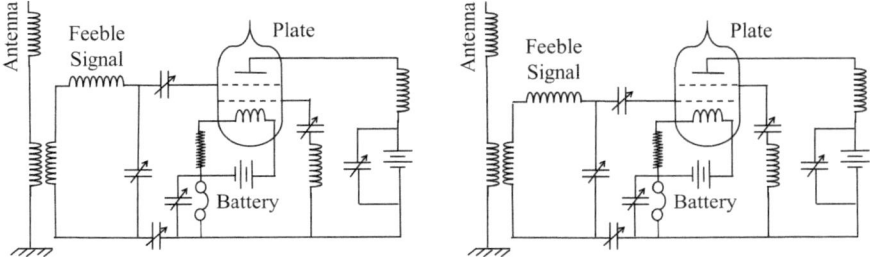

Fig. 6.12 Edwin Howard Armstrong's improved audion receiver circuits

Fig. 6.13 Edwin Howard
Armstrong's patented
regenerative circuit

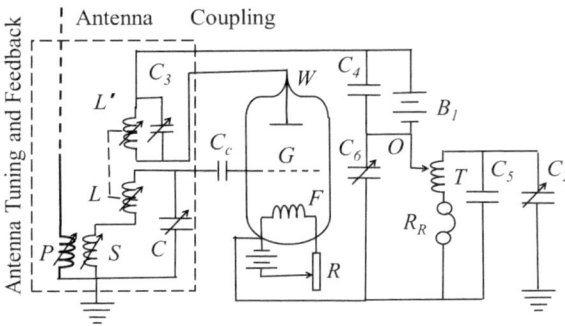

Edwin Howard Armstrong invented regeneration principle useful for selective amplification as well as high frequency oscillations. On *January 31, 1914*, Edwin Armstrong arranged the demonstration of his regenerative radio receiver at Belmar (NJ), USA before David Sarnoff, Chief Inspector of the American Marconi Company, and finally, he was issued with a US Patent No. 1113149 on *October 6, 1914*, and one of the better performing circuit out of many circuits he proposed is shown in Fig. 6.13.

The circuit consists of three parts, grid circuit, plate circuit and filament circuit. Grid circuit that is tuned with an antenna circuit and also fed back from the plate circuit for regeneration. Plate circuit is connected to filament circuit through capacitor C_6, part of auto transformer T and receiver R_R shunted with capacitor C_5 while grid and plate circuits are connected using part of auto transformer T and the capacitor C_2.

Edwin Howard Armstrong devised a "Circuit for using audion as a generation of high frequency generator" on *March 13, 1914* as shown in Fig. 6.14.

In his paper "Some Recent Developments in the audion Receiver" to IRE, on *March 3, 1915* that was published in *September 1915* gave a full account of the functioning of his circuit. However, Lee de Forest changed his patent disclosure and filed a patent for audion with the oscillating feature on *September 1915*, that was sustained in the favour of Lee de Forest on *May 21, 1934* by the United States Supreme Court. Both Edwin Howard Armstrong and Lee de Forest legally fought

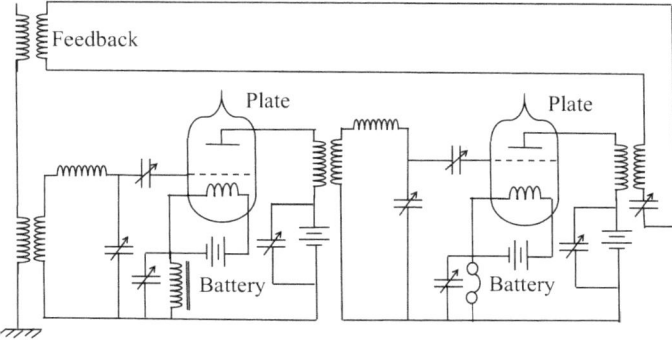

Fig. 6.14 Edwin Howard Armstrong's circuit for using audion as a generation of high frequency generator

Fig. 6.15 Edwin Howard Armstrong's regenerative circuit

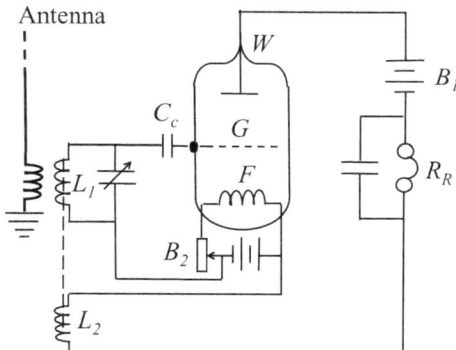

individually through corporations Westinghouse Electric and American Telephone and Telegraph, respectively, having purchased respective patent rights before US Supreme court's decision. However, scientific and engineering community believed on the contribution of Edwin Howard Armstrong and was in his favour, and believed that the court's decision was a mistake and the decision was based on lack of technical and scientific understanding. The these happenings, Edwin Howard Armstrong planned to return the medal of honour that was received by him in _1917_, however, Institute of Radio Engineers (IRE) decided to reaffirm the medal of honour to him in an IRE convention held at Philadelphia.

Edwin Howard Armstrong's regenerative radio circuit actually invented in _1914_, was published in his publication entitled "Some Recent Developments in the Audion Receiver" [139, 141, 142] of _March 3, 1915_ is shown in Fig. 6.15.

Inductors L_1 and L_2 are provides coupling for regenerative feedback from plate circuit to grid circuit.

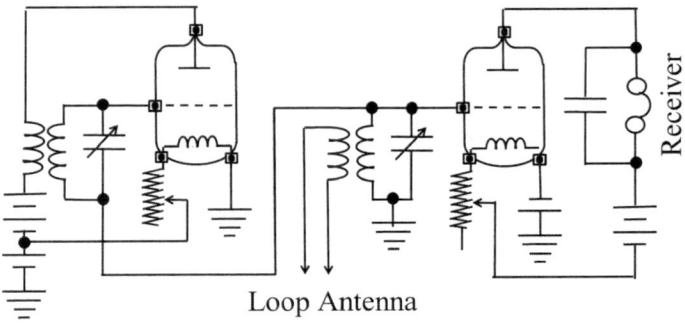

Loop Antenna

Fig. 6.16 Edwin Howard Armstrong's super regenerative circuit

6.3.2 Armstrong's Two Tube Super Regenerative Radio Receiver Circuit

In *1922*, Paul F. Godley, studied various types of regenerative circuits [141] of Edwin Howard Armstrong and discussed relative advantages and limitations of different types of variants from the viewpoints of operation and assembly. The two tube variant as shown in Fig. 6.16 was found to be the most suitable super regenerative circuit.

The first stage acts as an oscillator while the second stage is a regenerator that receives signal from the loop antenna and reproduces the received signal at its output using telephone receiver.

6.4 Patent Controversy Between Edwin Howard Armstrong and Lucien Lévy

The heterodyning was first experimented by Reginald Aubrey Fessenden in *1902* in which he could transmit waves of different periods and provided a suitable receiver to produce beat frequency at receiving end. Further implementation [143] added locally generated waves (local oscillator) provided substantial improvement that allowed tuning to different radio frequency carrier over which voice was made to ride. The practical demonstration was arranged between Fessenden's radio stations of US Navy at Arlington and Scout Cruiser; Salem and Birmingham and National Electric signalling Company in *1910*. In *1913*, John Hogan [144] provided further boost to the developments of this concept. In *1915*, Benjamin Liebowitz [145] put forward a theory heterodyne receivers.

In *1916*, A French officer Lucien Lévy in Telegraphie Militaire got an idea to modulate RF wave with supersonic wave modulated itself with audio signal. He tried to produce supersonic wave by heterodyning the received RF signal with the signal from the local oscillator and the resultant signal was selected by tuned circuits to produce the audio signal back. Lucien Lévy applied for two patents for his invention on *August 04, 1917*. Before the grant of the patents to him, AT&T acquired the patent rights, his claims substantially differed from those of Edwin Howard Armstrong's Patent 1,342885 granted on *June 08, 1920*, but as the advise of his counsel Levy broadened his claims covering the claims of Edwin Howard Armstrong. All the claims of Levy were supported by the Court of claims of District of Columbia [146] because his application was filed seven months before the application of Edwin Howard Armstrong and hence Lucien Lévy was awarded US Patent No. 1,734038 for his invention called "Electrical Transmission of Energy" on *August 12, 1918*. His patent covered heterodyning, amplitude modulation and frequency modulation aspects.

In *1916*, Edwin Howard Armstrong started experimenting with selective amplification using regeneration that laid to an invention of superheterodyne principle [147]. It takes the advantages of both regeneration and heterodyne principles to provide sensitive tuning of weak signals. This resulted in a work that was presented before IRE *Oct 4, 1916* in the form of his paper entitled "A Study of Heterodyne Amplification by the Electron Relay" which was published on *April 5, 1917*. In this paper, he describes self heterodyne circuits of regenerative type. Self heterodyning means amplification occurs at selective frequencies or both amplification and selection of frequency occurs simultaneously. In early part of *1917*, Edwin Howard Armstrong carefully studied heterodyne phenomenon. The general belief at that time was higher wavelength has a better range of coverage. Edwin Howard Armstrong's idea was to convert higher incoming frequency to lower that can be readily amplifiable. After extensive experimentation, an eight tube superheterodyne receiver was constructed having rectifier tube, three stages of intermediate amplifiers, second stage of rectifier and detector, two stages of audio amplifier and heterodyne oscillator.

In *1918* Edwin Howard Armstrong's superheterodyne principle [148, 149] that combines regeneration and heterodyne principles to provide sensitive tuning of weak signals. The renegation provides efficient method of amplification while heterodyning is provided by tuning an antenna to different frequencies related to different radio stations. These deadly combinations provide a wide base for very effective means of designing and deploying sensitive and selective radio receivers. The technical details of this invention were made public in *1919*.

On *February 1921*, Edwin Howard Armstrong published a paper entitled "A New System of Short Wave Amplification" [150] describes three types of practical circuits those cover amplification of audio frequency current after rectification, amplification of audio frequency current before rectification and application of heterodyne principle to increase the efficiency of rectification. In the process of amplification after rectification creates major problems in the performance of both amplification and rectification stages. The rectifier output is proportional to the square of input electromotive force or emf that makes it poorer. While amplifier also amplifies residual noise. In the second method, that is amplification is done such that signals can be dealt efficiently by the rectification and is efficiently applied for long waves (below 600 ms wavelengths) and are successfully applied for resistive, inductive and capacitive coupling. The third method greatly improves the efficiency of rectification when

Fig. 6.17 Edwin Howard Armstrong's signal detection and detected signal amplifier circuit

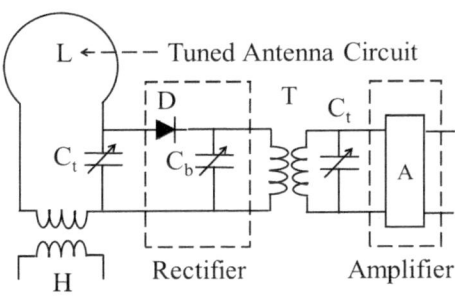

Fig. 6.18 Edwin Howard Armstrong's RF signal detection and RF amplifier circuit

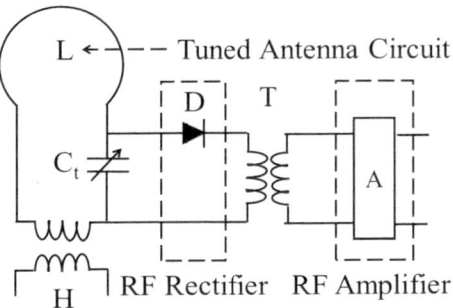

the signals are weak and are not efficient for long waves below 600 ms wavelengths. In spite of some problems, these direct solutions, that is, amplification after rectification or rectification before amplification was successfully applied by employing radio frequency amplifiers to cover a range of 300 m to 800 m wavelengths. This type of solution was secured by H.J. Round [151] by constructing tubes with small capacity without increasing their internal resistance beyond normal values and coupling tubes by means of transformers that provided high resistance to prevent oscillations at resonant frequency. Marius C. A Latour [152], instead of making modifications in tubes, introduced transformers using iron core transformers for coupling the tubes to solve the problem of oscillations at resonance frequency.

The circuit shown in Fig. 6.17 separates both RF frequency and signal that overrides RF carrier frequency, RF carrier is bypassed while detected signal without RF carrier is processed. The circuit consists of tuned antenna circuit consisting of loop antenna L and tuning capacitor C_t and heterodyne signal H, rectifier circuit consists of diode D and bypass capacitor C_b that bypasses RF frequency allowing only detected signal to be amplified by an amplifying stage and amplifier A for a detected signal.

The circuit shown in Fig. 6.18 processes both RF carrier together with a signal that overrides over RF carrier and hence requires RF amplifier capable of amplifying high frequencies. The circuit consists of tuned antenna circuit consisting of loop antenna L and tuning capacitor C_t, RF rectifier diode D and RF amplifier A.

If Frequency of 3 MHz is to be received, and beat frequency to be obtained as 0.1 MHz, then the frequency of the heterodyne signal (H) is set as either 2.9 MHz or 3.1 MHz. The combinations of various stages of amplification were employed by him.

6.5 Regulation of Wireless Audio Broadcasting and Its Licensing Leading to Organised Broadcasting

From the year *1912*, Department of Commerce, United States, started implementing licensing policies, the radio operators now required the licences for sending the signals on the air for communication or broadcasting entertainment programmes, and while enforcing these licensing policies, existing stations (particularly in Pittsburgh, KYW in Philadelphia and WOW in Omaha) were allowed to use their previously assigned call signs.

Charles D. Herrold received licences 6XE and 6XF (*December 1915*) for mobile transmission whose operations were interrupted during World War I, and only a few like Westinghouse Electric Corporation were allowed to operate.

6.6 Wireless Radio Systems During World War I

During World War I the focus was mostly shifted to wired and wireless telecommunication systems for defence applications like spotting of enemy portions, tapping of messages and making the systems portable, handy and light weight. The years starting from *28 July 1914* to *11 November 1918*, was a period of conflicts [153] leading to World War I. World War I was triggered following assassination of Archduke Franz Ferdinand and his wife on *28 June 1914* and was a conflict between central powers and allies as shown in Fig. 6.19.

During this period, Telecommunications went through a sea change that stressed on the implementation of many technologies, applications and inter operative network interconnections and tilted the interest in tele-net [154]. Allied forces were able to intercept the clear wireless messages of advancing German forces and virtually allied forces fought mobile war. Allied forces connected its forward positions to the headquarters using wired complex circuits connected using thousands of miles of copper core cables laid in underground trenches adopting telegraphy and telephony techniques avoiding plain language to keep German forces in dark about their plans. However, both sides started exploiting the vulnerability of each other.

In *1915*, a small number of trench sets were used but were not so successful and the enemy can tap the messages. The trench telephone set being used in trenches during world war I is shown in Fig. 6.20a, while trench set instrument is shown in Fig. 6.20b and c shows Marconi-Bellini-Tosi Wireless Direction Finder System

Fig. 6.19 World War I between central power and allies *Credit* United States Military Academy Department of History

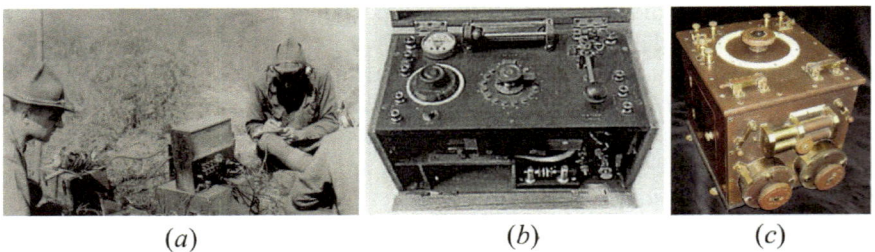

(*a*)   (*b*)   (*c*)

Fig. 6.20 a Receiving set for trench radio, with amplifier, receiving message. Signal School 1st Division, France, in World War I, Credit: DPLA and Hillary Brady **b** Trench Set, Spark with its spark gap clearly visible, Credit: The South African Military History Society and **c** Wireless Direction Finder—Marconi-Bellini-Tosi System Credit: CC0 1.0, Officine Radiotelegrafiche Marconi, |1910–1915|, Collezione di telecomunicazioni del Museo Nazionale della Scienza e della Tecnologia Leonardo da Vinci

During _mid-1915_ the Germans ahead in extracting information about plans of Allied forces. At end of _1915_, it was suspected that induction of lines was responsible for stealing of information. Allied forces after knowing the cause, introduced twisted pairs of wires instead of earth return. The ultimate solution to the problem was found

Fig. 6.21 Fuller Phone Mk IV, Courtesy Wireless for the warrior (WFTW)

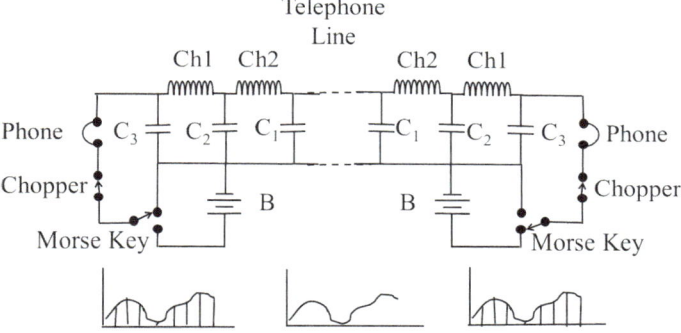

Fig. 6.22 Fuller Phone circuit

by Algernon Clement Fuller, member of IEE and Maj. Gen, Royal Engineer Signals, devised phone called "Fuller Phone" [155–157] as shown in Fig. 6.21.

Initially it was Morse coded telegraph transmitting coded messages using single wire that was modified for voice signals. The Fuller phone circuit is shown in Fig. 6.22.

Since communication was an important aspect in military operations, when military troops were advancing the efforts were also directed towards to lay down the telephone lines. The firing between the troops was also responsible for damaging the telephone lines, and hence many interconnections were provided for redundancy to continue with the communication and networks in the form of a matrix (or ladder) were being used. The tanks were also dedicated specifically for laying the cables in battlefield.

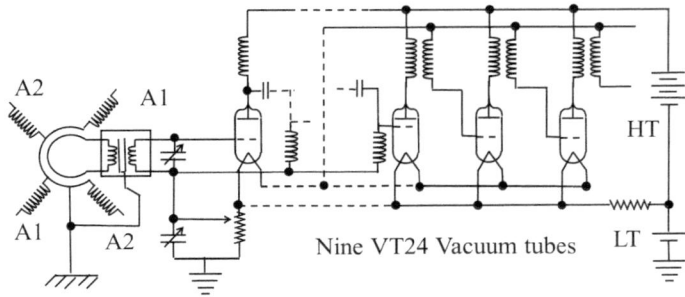

Fig. 6.23 Henry J. Round's direction finding amplifier

In *1916*, Marconi-Bellini-Tosi Direction finder was used to detect the positions of enemy wireless positions as shown in Fig. 6.20c. It was primarily used by the Royal Navy and British Army for spotting German submarines. During this time planes had no form of communications and heavily relied on signalling using tilting of wings, waving the flags and dropping of paper messages. The communication during that period heavily relied on spark gap mechanism, and hence was not suitable for aircraft communication for the safety of plane carrying combustible fuel. This scenario was changed after the invention of radio valves and in *1916*, Royal Flying Corps started developing radio telephony sets so that telephonic conversation can be established between aircraft and ground stations to tell ground stations where exactly shells are landing.

Captain Henry Joseph Round from Marconi Company took the assignment for British Military Intelligence since *1902*, developed wireless set weighing 25~26 Kg, was easy to carry it by balloons for intelligence gathering. Henry. J. Round's single man affair of wireless innovation provided great transformation for military intelligence. Further, valves provided more compactness in the construction of wireless sets to be suitably used in aircrafts for collecting intelligence from skies and sending the coded information to ground artillery. In *1917*, Henry. J. Round design his own vacuum tube valve called V24 with very low inter-electrode capacitance helped in reducing inter-electrode interference. He designed a multistage (9 tubes) direction finding amplifier [151] using V24 tubes with a gain of 2000 for army intelligence. The antenna used was a single turn antenna to which each pairs A1 coils and A2 coils were connected at 90°. Some direction finding amplifiers used as many as 130 tubes. The Rounds direction finding amplifier is shown in Fig. 6.23.

The German Foreign Minister Arthur Zimmermann's telegram to German Ambassador in Mexico City that Germans will help Mexico to reclaim its territory in possession of United States was intercepted by Naval intelligence of UK in *January 19, 1917*. Due to this act of Arthur Zimmermann, US dragged in to world war I. In United States, manual switchboard operated civil telephone systems were extensively used for military communication purpose and women operators played a crucial role as switchboard operators during this world war.

(*a*) Front Panel view   (*b*) Interior view

Fig. 6.24 Marconi Type 106 crystal receiver, Courtesy by Howard Stone, stonevintageradio.com

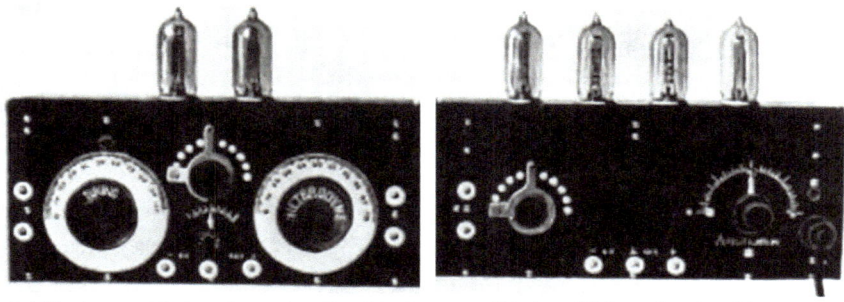

(a) Tuning and heterodyne circuit cabinet   (b) Amplifying circuit cabinet

Fig. 6.25 CC BY-SA 3.0, H. W. Houck |1920| H. W. Houck, "The Armstrong Super-Autodyne Amplifier, part 1" in Radio Amateur News, Experimenter Publishing Co., New York, Vol. 1, No. 8, February 1920, p. 403 on Google Books

American Marconi Company developed sophisticated crystal radio receiver (Marconi Type 106 crystal receiver) from *1915* to *1920* and was gone through many changes that was used in transatlantic communication in *1917*, is shown in Fig. 6.24.

6.6.1 Edwin Armstrong's Superheterodyne Receiver

As per the disclosures made by Edwin Armstrong in his letter to Major O.E. Buckley, written on *June 05, 1918* [158], it is apparent that one of the earliest and first form of superheterodyne radio receiver prototype was built by Edwin Armstrong at US Army Signal Corps laboratory in Paris [159] during World War 1 in *June 05 1918*. The photographs of superheterodyne receiver in the form pair of cabinets containing Tuning and heterodyne circuit and Amplifying circuit respectively is shown in Fig. 6.25.

The tuning and heterodyne circuit contained in a cabinet as shown in a photograph of Fig. 6.25a are shown in Fig. 6.26. The circuit consists of Antenna Tuner, Hetero-

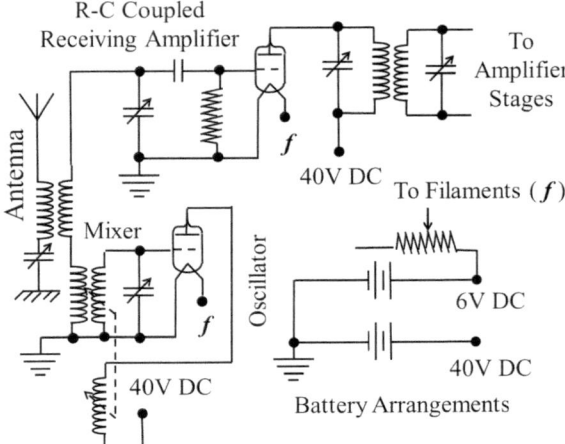

Fig. 6.26 Edwin Armstrong's Antenna Tuner, Heterodyne Mixer, Oscillator and RC Coupled Receiving Amplifier circuits contained in a cabinet as shown in Fig. 6.25a

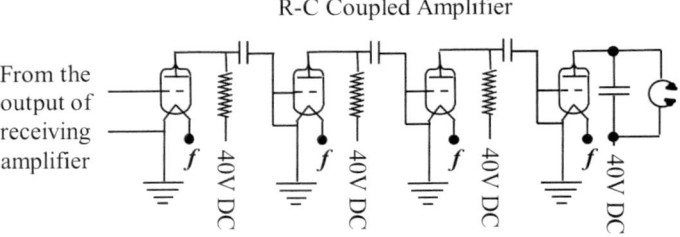

Fig. 6.27 The amplifier circuit of Edwin Armstrong's superheterodyne contained in a cabinet as shown in Fig. 6.25b

dyne Mixer, Oscillator and RC Coupled Receiving Amplifier circuits. The circuits of the cabinet as shown in the photograph of Fig. 6.25b are shown in Fig. 6.27.

By the year *1918*, Britishers mastered the art of aircraft to aircraft communication. During *1918*, communication was using continuous carrier wave radio sets, spark or loop sets which were fairly unsophisticated and primitive type.

On *11 November 1918*, a face saving state of peace agreement was reached between Germany and allied forces [160] called "Armistice of Compiégne", in which Germans agreed to the polices proposed by U.S. President Woodrow Wilson in his 14 point proposal prepared by French Marshal and Supreme Commander of the Allied Armies Ferdinand Foch. The German side was represented by Kaiser Wilhelm II, his Imperial Chancellor, Count Georg von Hertling at Imperial Army Headquarters, Spa, Belgium (Which was occupied by Germans) and Oberste Heeresleitung, a German Supreme Army Command.

6.6.2 State of Radio Broadcasting During World War I

In *1914*, at the University of Wisconsin, Madison, Professor Edward Bennett established a spark gap transmitter broadcasting station capable of transmitting Morse code transmission. During next year *1915*, this station equipment was transferred to Earle M. Terry. From *June 1915* U.S. Department of Commerce started regulating radio services, issued experimental license with the call sign 9XM used for P2P (point to point) communication with other stations. This station started transmitting weather reports using Morse code in *December 1, 1916*, using wavelength 1700 m (176 kHz). In *1917*, E. M. Terry started testing of transmission of phonographic records incorporating newer technologies using vacuum tubes.

In *January 19, 1917*, United States got involved in World War I due to German Foreign Minister Arthur Zimmermann's telegram. During *February, 1917*, Germans targeted and sank many US submarines. On *February 26, 1917*, the President Woodrow Wilson convened a joint congress session on for declaring war against Germany, however, many anti war senators consumed all of the congressional session and received great setback to his proposal and hence on *February 26,1917*, President Woodrow Wilson ordered to equip all merchant ships with military hardware, citing an old anti-piracy law in spite of opposition by some congress members.

President Woodrow Wilson again went before joint congressional session on *April 2, 1917* to declare a war against Germany which was supported by U.S. Senate on *April 4, 1917*.

With the entry of United States, for safety reasons, all amateur and commercial broadcasts were stopped abruptly on *April 7, 1917* by the order of US President in public interest and become an act of disloyalty, crime and the possession of radio equipment became illegal. The navy was highly involved in wartime radio and was allowed to screen all incoming and outgoing telegrams.

World War I ended with the signing of armistice by Germany with allies in favour allies on *November 11, 1918*. Although ban on civil radio operators was in force till *October 01, 1919*, the restrictions on private and amateur listening ended on *April 15, 1919*.

6.7 Wireless System Development Scenario After World War I

World War I triggered multifaceted developments in the area of wireless radio systems and these developments in wireless systems after World War I can be seen from various perspectives such as radio frequency bands, amateur wireless radio systems, wireless equipment industry, broadcasting and entertainment industry, licensing authority that also followed systematic approach. Besides this, attention was paid to efficiency, reliability and robustness of the apparatus.

6.7.1 Amateur Wireless Radio Communication Systems

Amateur systems flourished after World War I, the wireless radio kits were available for experimentations the led to drastic developments in wireless communication field.

6.7.1.1 Amateur Radio Call Sign Assignments in Europe

Various Europeans counties like United Kingdom (*1920*), France (*1921*), Luxembourg & Italy (*1923*), Finland (*1924*), Belgium & Germany (*1924*), Denmark (*1924*), Switzerland and Netherland (*1924*) received respective prefixes for call signs as (2,5,6), (8), (1), (3), (4), (7), (9) and (0) respectively.

6.7.1.2 Promotion of Amateur Wireless Radio Developments by American Radio Relay League (ARRL)

On *January 1921*, a group of amateurs published "Pacific Radio News" specifically dealing with amateur radio activities and in the same year amateurs organised continuous wave (CW) communication using tube transmitter across the Atlantic ocean working on wavelengths shorter than 200 m or at frequencies more than 1500 kHz. After some unsuccessful attempts American Radio Relay League (ARRL), a noncommercial national association for Amateur Radio in the US that was founded in *1914* by Hiram Percy Maxim, in its first "ARRL Convention" at Chicago decided to register the signals from American amateurs to Europe [161]. In this venture, the British amateurs played a crucial role, Mr. P. R. Coursey, Editor of Wireless World (London) representing British amateurs obtained permit form British Post office for conduction of such experiment and acted as referee throughout the experimentation. Before conducting the experiments, test were to be conducted on various equipments for the range of about 1000 miles overland, failing of which equipment was to be disqualified. From *November 1* to *5, 1921*, tests were conducted for which seventy-eight star stations of radio district were chosen and Paul M Godley's superheterodyne receiver with 3 foot loop antenna and other twenty-seven contestants were chosen. On *November 14, 1921*, Paul Godley sailed to Ardrossan, Scotland, on an ocean liner called "Aquantina" with state of art equipment to receive the amateurs signals from the United States. The schedules of test from *December 07–16, 1921* was worked out that was divided in to two slots from 7.0PM to 9.30 PM and 9.30 PM to 1.0 AM with each period of 15 min was allowed for each of the nine inspecting radio districts. These slots were rotated for each night for uniform distribution of time to each of nine inspecting radio districts. The first slot was free for all while the second slot was allowed to station qualified in preliminary tests. Each of the qualified stations were assigned with five letter secret code. In this experimentation, all great commercial stations have put the disposal of ARRL's experiment by the efforts of RCA's traffic manager Winterbottom.

On the *December 7, 1921*, the equipment was set up on the coast of Scotland. The observer appointed for this was D.E. Pearson, from Marconi Marine Communication Company. On the night of *December 9, 1921*, 1AAW in Roxbury, Mass., USA was heard at 12.50 am of next day of *December 10 1921*, after listening for the sparks, Paul Godley switched over for continuous wave reception, Godley picked up a station with call sign 1BCG Greenwich, Connecticut, USA, operating on 230 m that was interfered by the bothersome harmonic of a station operating at Clifden, Ireland that was working at the distance of 150 miles. In an effort to remove the interference from this station, many time the contact with the station 1BCG was lost, and hence efforts to connect 1BCG were terminated at 1.33 AM (*December 10, 1921*) and tried to reconnect, it called "PF" several times which was fading after every second and at 1.59 am, 1BCG called station 2BGM saying "Phone us now" and shut off and nothing was heard from 1BCG. The signals from 1BCG were seemed to be loud enough to be heard to a distance of 400 feet from the tent and later on it was learned that the input was 990 W and seemed to be the most reliable signal received. Other stations similar in the strength were 1ARY, 2FD and 2FP while 1BDT was not as strong as 1BCG, it equalled the record of 1BCG in the steadiness of the signal. On following day that is, on December 11, 1921 at 2.30 AM, Paul Godley received a complete twelve word message from 1BCG received at Ardrossan, Scotland that was signed by Burghard, Inman, Grinan, Armstrong, Amy and Cronkhite. This message proved to be the first transatlantic message sent via amateur radio. The subsequent stations logged up to 3.30 AM were continuous wave stations: 3XM, 1BKA, 1XM, 1BCG, 2EH, 2FP, 2ARY, 2AJW, 1ARY, 1RZ and spark stations:1BDT, 3FB, and 2EL.

The 9 tube superheterodyne used in this experiment consists of four types of circuits (a) A circuit that implements wave frequency tuning, regenerative circuit at wave frequency, oscillator and mixer circuit using two valve tubes, (b) five stages of rf amplifiers implemented using five valve tubes, (c) Detector implemented using one valve tube and (d) Audio amplifier stage implemented using one valve tube. The initial stage circuits are shown in Fig. 6.28.

The circuits related to final stages called last stage rf amplifier circuit, detector and audio frequency circuits are shown in Fig. 6.29.

6.7.2 First Two Way Transatlantic Communication

During *1922*, it was noticed that the British has difficulties to hear US stations, amateur Leon Deloy (8AB), improved his wireless radio system based on inventions of Forest and Armstrong, and the contact between Leon Deloy was made with his counterparts John L Reinartz (1XAM) and Fred Schell (1MO) [98] in USA on November 27, 1923 using wave of 110 m (2720 kHz) as shown in Fig. 6.30.

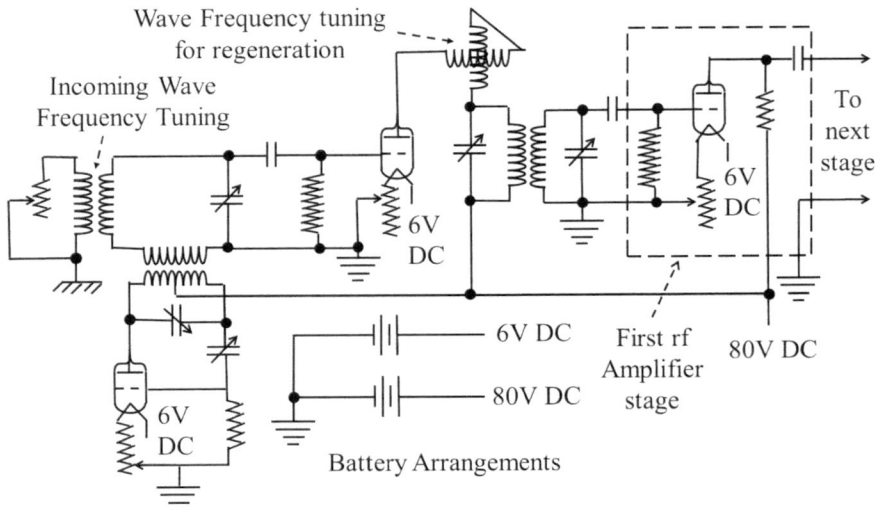

Fig. 6.28 Antenna tuning, heterodyne mixing, oscillator, regeneration tuning and first rf amplifier stage circuits of superheterodyne receiver used by amateurs in transatlantic communication

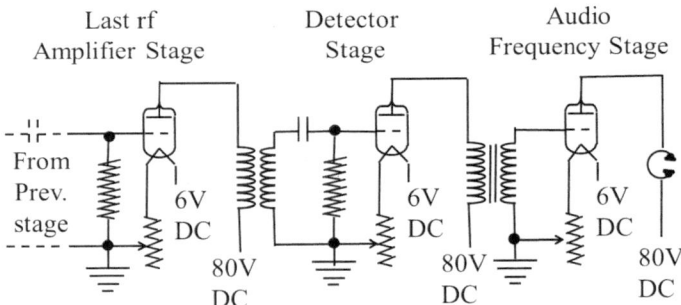

Fig. 6.29 Last stage rf amplifier circuit, detector and audio frequency circuits of superheterodyne receiver used by amateurs in transatlantic communication

Fig. 6.30 The wireless transmitter (1MO) used in first two way transatlantic communication on a wavelength of 110 m that used vacuum tube technology, Courtesy QST © Copyright 2010 American Radio Relay League, Inc.

6.7.3 Status of Wireless Broadcasting Systems After World War I

After World War I, on *October 01, 1919* licences of radio operators were resumed in the United States of America. Broadcaster exploited the tremendous advances in vacuum tubes technologies, crystal detector technologies in wireless systems. These technologies set to enter for the improvements in wireless transmitters and receivers.

It was the time when point to point wireless communication using telegraphy or telephony systems were still in use. The broadcasting, as well as radio receiver industries, flourished together to their mutual but complementary benefits.

Due to the spectacular growth in broadcasting during the decade of *1920s*, the inevitable and insatiable demands were created for radio receivers. This resulted in rapid growth of radio industry.

Till *1923*, bulky "breadboard radios" consisting of components like rheostat, vacuum tubes, variable condensers and large coils flourished.

In the beginning of *1923*, breadboard radios started disappearing and after the invention paper cone speaker by Peter Jenson, radio receivers with a speaker with ornate metallic enclosure and stylish wooden cabinet enclosed radios started appearing in the market.

During *1923* to *1927*, radio receivers increased many folds those incorporated many modifications and improvements like single knob tuning that simplified the complicated receiver designs.

6.7.3.1 Spread of Wireless Broadcasting Stations After World War I

An Engineer at Westinghouse, Frank Conrad, attracted by broadcasting technology, built a radio transmitter on the second floor of Wilkinsburg garage was registered as 8XK call sign that was a forerunner of KDKA. Frank Conrad, was responsible for making transmission of music and voice from two call sign signals of 8XK and 8YK since *1916*. After world war I, the ban on civil radio stations was lifted and 8XK became KDKA of Westinghouse Electric Corporation of East Pittsburgh, Pennsylvania in *October 27, 1920* when was first limited commercial licensed Radio Station in the United States of America. Frank Conrad, an executive at Westinghouse, broadcasted the records of his friend that sent a strong message about how radio broadcast is a powerful medium for masses and from here onwards the idea of public radio broadcast took the shape. On the election night, KDKA started its historic broadcast at 6 pm on Tuesday, *November 2, 1920*. For the benefit of their listeners, the Westinghouse in USA, started manufacturing radio receivers sets for listening broadcasted programmes. KDKA charged no fees to its users and did not carry any commercial paid advertisements. These activities were financed through the sale of radio receivers. These developments sparked the deployment of broadcast radio systems worldwide.

Wisconsin University's Experimental call sign license 9XM was restored and work of transmission of weather forecast broadcast was resumed on *January 1920*, however, broadcast service for telephony could not be upgraded due to lack of vacuum tubes, and finally, the equipment was ready by *January 3, 1921* and this was the first time when voice-based weather broadcast was made.

The major drawbacks of radio wireless systems was that anybody can listen to transmitted signals leading to unsecured communication, an idea to convert this drawback into advantage was in the mind of a man called David Sarnoff [162] of American Marconi Company, however his pre-World War I idea seemed to premature due engagement of industry to World War I. David Sarnoff's idea of transmitting music programmes for benefit of many was realised in *15 January 1920*, when a 6 kW transmitter was installed at Marconi Works at Chelmsford for testing the reception performance of radio receivers.

In early *1920*, British Thomson Houston started manufacturing radio receivers so that audience can access transmitted programmes. This was the time when the term "broadcast" really came into existence. In Britain, the first experimental public broadcast began in *June 15, 1920*, from Marconi's factory in Writtle near Chelmsford. A singing recital by Dame Nellie Melba was broadcasted using 15W Marconi's telephone transmitter and signal was received at many places.

In Holland, Hanso Henricus Schotanus á Steringa Iderza was perusing its commercial interests by starting sell of radio sets, and on *November 6, 1919* he started transmission under call sign PCGG and by *November 1921*, he was able to convert this in to a full broadcast licence.

The first publicised broadcast by Marconi Company at its factory premises at Writtle near Chelmsford, Essex, UK was commenced on *June 15, 1920* using two 450 feet high broadcast masts. In *1921*, Marconi Company was permitted to broadcast regular music and entertainment programmes from its transmitter at Writtle.

Westinghouse again establishes the radio station licensed with call sign WJZ on *June 1, 1921* in Newark, New Jersey, but it could no go on air till *October 01, 1921*. However due to its location, it became most important radio station of Westinghouse. It was the first radio station in the vicinity of New York.

Westinghouse established its yet another radio station with call sign WBZ at Springfield, Boston, MA, USA, and was the first radio station to receive commercial licence and started its broadcast on *September 19, 1921*. Soon it opened its studio at Hotel Kimball.

RCA established its first experimental commercial Radio Station with call sign WDY that was located at General Electric's Aldene plant in Roselle Park, New Jersey licensed on *September 19, 1921* and did not go in operation till *December 14, 1921* and started its commercial broadcasting operations from *December 15, 1021* using 360 ms wave or 833 kHz. WDY's 500-watt transmitter was built by GE engineers. Although WDY's licence was issued before WJZ, its broadcasting operations were commenced two months after the commencements of broadcasting operations of WJZ. RCA made its decision to shut down and WDY made its final broadcast on *February 17, 1922* and finally its licence was deleted on *February 20, 1923*, however WJZ still remained in the race. The opening of many competing radio stations like

[163] General Electric's flagship radio station WGY (**W**ireless **G**eneral Electric in Schenectad**Y**) New York was in air on *February 20, 1922* operating on 360 ms wavelength(833 kHz); WOR in *February 22, 1922* at AM frequency 710 kHz; WSB ("**W**elcome **S**outh, **B**rother") in *March 15, 1922* operating on 750 kHz; WHN in *March 18, 1922*, at AM frequency 833 kHz; WAAM operating on 349 m/859 kHz in *April 10, 1922*; WDAP, *May 19, 1922* operating on 720 kHz forced WJZ to voluntary operate on part time basis.

At 4:30 on the afternoon of *November 11, 1921*, a new broadcasting station, the first station in the Chicago area, KYW, went on the air with a program broadcast from the stage of the Chicago Civic Auditorium. In the same year, Herrold obtained licence for KQW which become KCBS, a state owned radio station CBS in San Francisco. It was the first time on *December 07, 1921*, a group of six amateurs chosen by the Radio Club of America built shortwave station at Greenwich, Connecticut (call sign 1BCG), sent message across Atlantic ocean using short wave that was also picked up in Ardrossan, Scotland by Paul Godley (2ZE), that proved the utility of short waves having wavelengths below 200 m (1498.96229 KHz or 1.4989 MHz) that opened a door for short wave communication.

In *February 1922*, Marconi Company got call sign 2MT for station at Writtle, Clemsford, Essex, UK and started broadcasting songs, competitions and even a lonely hearts club and by *May 1922*, Marconi Company got another licence to transmit from Marconi's London studio at Marconi House Building in the Strands, London under call sign 2LO.

WEAF was the first commercially licensed radio station in New York City owned by AT&T Western Electric. Western Electric AT&T Fone (WEAF) was a competing station with WJZ. WEAF acted as a research centre and also maintained a regular schedule of its commercially sponsored programmes. It also played a leading role in the networking of broadcasting stations. It successfully connected AT&T's WCAP station at Washington DC and WJAR station at a province in Rode island.

A flagship AM station WNBC of NBC Radio Network was the commercially licensed radio station in New York City operating on 660 kHz and 50,000 W. Looking at the networking success of WEAF, first time on *March 02, 1922* WNBC signed as WEAF, owned by AT&T Western Electric.

The sale of radio receivers by broadcaster was justified to recover broadcasting expenses, and according to National Association of Broadcasters, there were 60,000 home users in *1922* that rose to 10 million in *1929*. During *1922*, there was an increasing influence of commercialisation on broadcasting services faced with stiff opposition that was evident from Radio Conference on *1922* in Washington DC, wherein Herbert Hoover, Secretary of Commerce in charge of broadcast regulation, expressed concern over commercialisation and disapproved it, fearing that broadcasting business will be drowned in advertisement and chatter.

At this time, licence office was flooded by huge number of applications and the situation became chaotic and hence General Post Office (GPO), a licensing authority proposed a single licence to a company owned by a consortium of leading wireless receiver manufacturers became British Broadcasting Company (BBC) founded on 18 *October 1922*. On *November 14, 1922*, British Broadcasting Company (BBC) was inaugurated and started its broadcasting from Marconi House Building in the Strands, London under call sign 2LO. A Scottish John Reith was appointed as general manager of BBC on *December 1922* when official daily broadcast of British Broadcasting Company (BBC) [164] started.

In early *1922*, American Telephone and Telegraph Company (AT&T) [165] tried to explore possibilities of charging fees for airing commercial advertisements on their stations, however due to fear of legal action, and was persuaded to sell its stations to Radio Corporation of America (RCA) to leave broadcasting services and in exchange AT&T was offered exclusive rights to connection access and link local station of NBC network. RCA now started to contribute 50% to WJZ.

On *January 4, 1923*, linking between flagship station of AT&T WEAF with the most legendary radio station in the history of Boston called WNAC was achieved.

In *May 14, 1923*, the radio station WJZ was acquired by Radio Corporation of America, and hence WJZ faculty was moved to New York City and commenced its operations on *May 15, 1923* by inheriting call sign WJZ. Till now, radio stations were operating on radio waves ranging from 360 ms (832.75 kHz) to 400 m (749.48 kHz) and on *May 15, 1923* AM broadcast band from 550 to 1350 kHz was opened that was further split into 550–1040 kHz for class B stations and 1050–1350 kHz for class A stations and these new assignments came in force for Class B stations like KDKA and WJZ started operating on 920 kHz and 660 kHz, respectively.

WGY at New York, which went in air on *February 20, 1922*, operating on 360 ms wavelength (833 kHz), form *May 15, 1923*, onwards, the station started operating on 790 kHz that was previously being used by WHAZ which was migrated to 1300 kHz.

Also during *1923*, regular broadcasts were started in Czechoslovakia and Germany on May 18 and *October 29th 1923* respectively while Australia starts its first licensed radio station 2SB begins its operation on *November 13, 1923* at Sydney. On *December 31, 1923* KDKA conducts its transcontinental voice transmission with a radio station in Manchester, UK.

The radio stations WJAR, WCAP, and WEAF broadcasted Woodrow Wilson's funeral on *February 6, 1924* and also for the first time these stations started sponsored programmes under the sponsorship of the National Carbon Company. On *February 24, 1924*, First US President Calvin Coolidge to broadcast a program on Washington's Birthday from the White House. On *June 1, 1924*, WDAP becomes WGN (World's Greatest Newspaper) operating on 720 kHz. From *June 24– July 9, 1924*, 18 stations carry out the WEAF broadcasts related to a Democratic convention in New York. During *November 1924*, the upper limit of broadcast band 1350 kHz was further extended to 1500 kHz yielding an additional 15 class A frequencies for use.

By this time, AT&T had a network of 21 stations that was inaugurated by US President Calvin Coolidge on *March 04, 1925* with the estimated audience of 22,800,000.

On *July 29, 1925*, first high power transmitter 2XAG, operating on 50 kW at Schenectady NY become operational. The first use of crystal in radio station WGY in *October 10, 1925*.

History of broadcasting in Asia was created when telegraph department of Ceylon started experimental broadcasting in *1923*, now known as Sri Lanka Broadcasting Corporation which was actually launched officially on *December 16, 1925*.

On *July 8, 1926*. Attorney General renders a decision that the Department of Commerce has no jurisdiction over radio frequencies.

National Broadcasting Corporation (NBC) Network was jointly formed by GE, RCA, and Westinghouse on *September 9, 1926* and its operations were started from the studios of AT&T's WEAF radio station in New York City on *November 15, 1926*. NBC was a combined group of nineteen scattered stations connected by telephone wire circuits equivalent to 3500 miles. WSB ("Welcome South, Brother") operating since *1922* got affiliated to NBC on *January 9, 1927*. With its innovative techniques, it became the most popular and oldest major broadcasting network.

6.7.3.2 Spread of Wireless Amateur Broadcasting Stations After World War I

In *1922*, radio amateurs were operating in the small band from 1500 to 2000 kHz with small transmitter power as compared to public and commercial services. However, to compensate for low power that was posing a problem in achieving longer distances, amateurs used their skills to design antennas and tuned circuits at higher frequencies to achieve radio communication at the distance of thousands of kilometers and the reality was two-way communication between North American and European amateurs. There was a need to extend the Band of 1,500–2,000 kHz to a much more higher frequency due to congestion and experiment of 1MO station in USA with its European counterpart 2720 kHz or ∼3 MHz showed better performance.

In *1923*, US Commerce Department responsible for issuing broadcasting licence, realised that great demand for issuing of licences can not be practically satisfied for every application for licensing, hence secretary, Department of Commerce, Herbert Hoover worked out a medium wave frequency assignment plan for broadcasting stations and set aside 81 frequencies separated by 10 kHz starting from 550 to 1350 kHz. The higher end, that is 1350 khz was later on subsequently changed in later years to 1500 kHz, 1600 kHz and 1700 kHz, respectively.

In *July 1923*, Charles Franklin sent messages from large antenna in Poldhu, England to Marconi's at the Capo Verde Island (West Africa), using 25 Kilowatt transmitter operating on 3 MHz. Marconi was highly encouraged by the experiments of Charles Franklin and hence he contacted British Postmaster General for negotiations for the establishment of network of shortwave radio stations ("Imperial Wireless Network") covering England, South Africa, India, Australia, and North America.

In another attempt on *September 1924*, Charles Franklin made a contact with the yacht of Marconi at the Beirut Port, Lebanon and continued his improvements in antennas for shortwave communication.

In *September 1924* amateurs in California established a contact for 90 min with amateurs in New Zealand, owing to these happenings, third Radio Convention in the

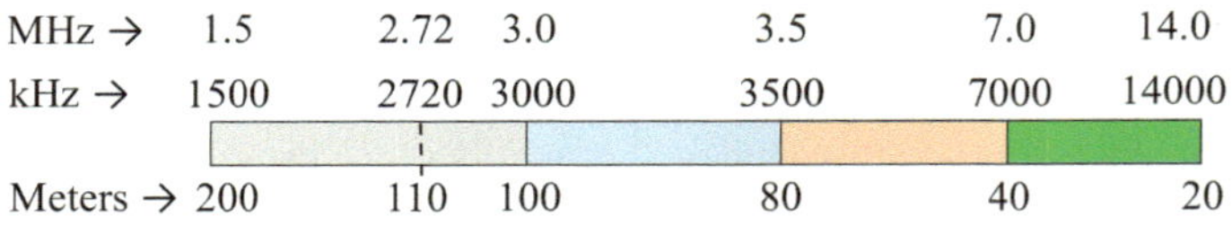

MHz →   1.5   2.72   3.0   3.5   7.0   14.0

kHz →   1500   2720   3000   3500   7000   14000

Meters → 200   110   100   80   40   20

Fig. 6.31 Allocation of short wave frequencies to amateurs as on September 1924

Table 6.4 First broadcasting stations in various countries

Country	Date of Start	Remarks
America	1912	2XG New York
America	Nov 02,1920	KDKA Pittsburgh
Britain	February 1922	2MT Writtle
Canada	1919	XWA, Montreal
Australia	1923	2SB, Sydney
India	June,1923	Radio club Bombay
Germany	June,1923	LP Berlin
France	January 20, 1923	Radio PTT Paris
Cuba	October 10, 1922	PWX Havana
Japan	March 1925	JOAK, Tokyo
Mexico	November 1923	CYL, Mexico City
Philippine	1927	KZRM Manila
Sri Lanka	December 16, 1925	Colombo

USA allocated new radio frequency bands that is 3.5 MHz (80 m), 7 MHz (40 m) and 14 MHz (20 m) are shown in Fig. 6.31.

In *1926*, Radio Corporation of America (RCA), initiated transcontinental radio services through its National Broadcasting Corporation (NBC), its subsidiary Red Network and Blue Network. The Columbia Broadcasting System (CBS) was founded by William S. Paley in *September 18, 1927*.

6.7.3.3   Staring of First Radio Station Broadcasts in Various Countries

The progress of setting up of first radio stations in various countries is shown in Table 6.4.

6.7.4   Crystal Radio Receiver Developments

After World War I, crystal detector-based wireless radio sets were inductively or capacitively tuned radio revivers, and hence these wireless radio sets mostly dom-

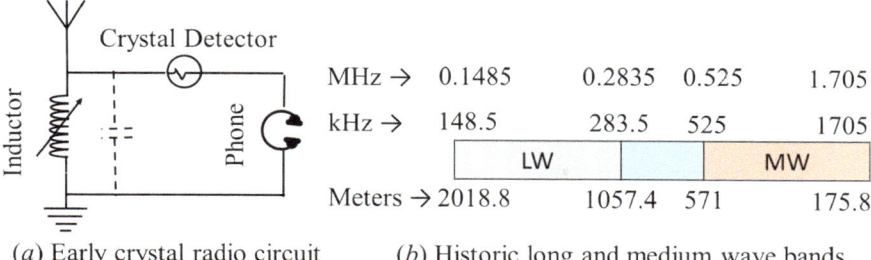

(*a*) Early crystal radio circuit   (*b*) Historic long and medium wave bands

Fig. 6.32 Early crystal radio circuit and their bands of operation

inated by tuning, crystal detection and phone receiver that imparted simplicity to them and can be operated without much complications. Tuned circuits supposed to be constructed using parallel circuit consisting of inductor and capacitor, however these early circuits were consisting of only four components called antenna, inductor, crystal and phone receiver. The tuned circuit was constituted of inductor and capacitor formed by antenna with ground plane is shown in Fig. 6.32.

These radios were historically designed to tune to long waves (148.5 to 283.5 kHz) or medium waves (525 to 1705 kHz) depending on the value of inductor by using a coil with tapings or continuously varying inductor using slider arrangement. There were slight changes in the definitions of medium wave bands depending on the country and region.

In *1920*, A.C Gillbert, a toy maker, New Haven, Connecticut, marketed radio set model 4006 based on primitive crystal detector, while in the year *1922*, it marketed crystal radio set model 4016 [166].

The three radio wireless companies played crucial role in promoting the wireless radio sets [167] are Wireless Speciality Apparatus Company (Estd. *1906*), Radio Corporation of America (Estd. *1920*) and Ajax Manufacturing Company (Estd. *1921*) which was incorporated under name Fillmore Manufacturing Company in *1925*. In the beginning of *1920s*, that is, during *1921*, factory made radio sets were expensive and hence these companies provided cheaper kits that helped amateurs to spread the radio market. Fillmore Manufacturing Company was most successful in dealing with crystal sets and components like detectors, crystals and hardware. Fillmore Manufacturing Company was most successful in selling ready to use crystal detector assemblies enclosed in a glass enclosure to various companies manufacturing wireless crystal radio receiver sets for individual users. Many radio receiver manufacturing companies were then able to sell radio receivers at lower cost, besides this, amateurs were able to construct radio set themselves.

On *March 16, 1922*, Department of Commerce Bureau of Standards Washington, USA [168], released the construction and working of a simple radio receiving equipment circuit as shown in Fig. 6.33.

In *April 1922*, James Leo McLaughlin, published a circuit in "Science and Invention", won the first prize of $100.00, and its implementation is shown in Fig. 6.34.

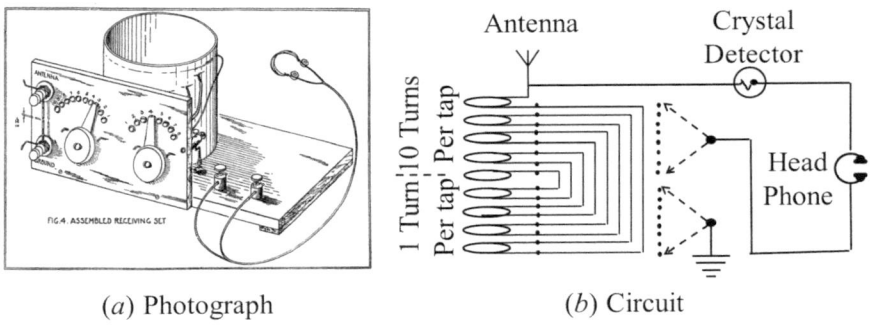

(a) Photograph   (b) Circuit

Fig. 6.33 Simple circuit of wireless crystal radio set by Department of Commerce Bureau of Standards Washington, USA., Courtesy of HathiTrust

Fig. 6.34 James Leo McLaughlin's award winning circuit of April 1922

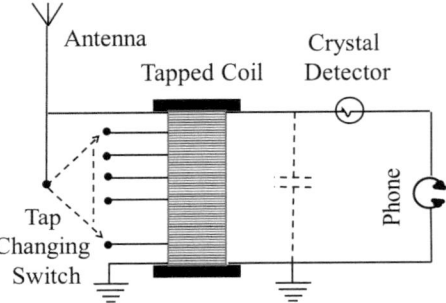

6.7.5 Taking over of Marconi Wireless Telegraph Company of America by GE

During and at the end of World War I, Marconi Wireless Telegraph Company of America had an excellent standing in acquired patent rights to give it a leading role in the United States of America. A controversial act passed by the US congress in *June 25, 1910* provided the US navy the powers to disregard patent rights when awarding the contracts, fearing the legal complications, Marconi Wireless Telegraph Company of America agreed to sell its assets to General Electric. This provided a solid foundation for General Electric to launch its new subsidiary company called Radio Corporation of America on *17 October 1919*.

In *October 22, 1919*, Assets of Marconi Wireless Telegraph Company of America (also called as American Marconi) were acquired by General Electric [169] The new subsidiary company called Radio Corporation of America, played a crucial role in fueling the growth of radio receiver market. The cross licensing agreement was signed between Radio Corporation of America and General Electric Company on *November 20, 1919* [170] that entitled Radio Corporation of America to acquire all the right over all the patents of General Electric Company including recently acquired

Fig. 6.35 Radiola-I Crystal radio set, Courtesy Antique-Radios.net: Information site for antique radio collectors

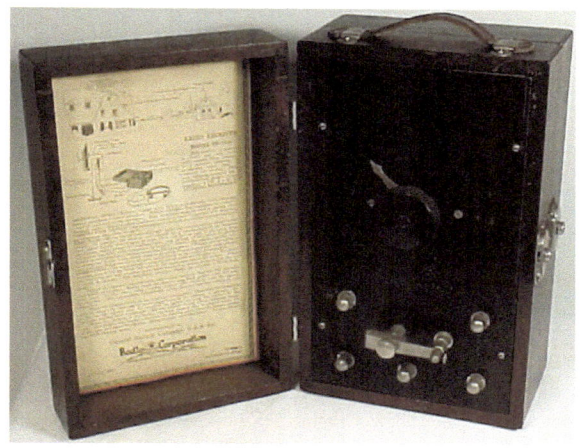

patents of Marconi Wireless Telegraph Company of America held by General Electric Company. With this agreement, RCA took over the ownership and management of all high power wireless stations that were previously owned by Marconi Wireless Telegraph Company in America. RCA also took over ownership of various activities of American Marconi Company that included branches of Marconi Telegraph Cable Company Inc. at New York, Massachusetts and Illinois; Wireless Press, New York; and Pan- American Wireless Telegraph & Telephone Company of Delaware. In *May 4, 1920*, a powerful and potent American Telephone & Telegraph Company having a vast network of landline had an edge over Westinghouse makes an alliance with RCA. However, in this alliance, Westinghouse could not grab the opportunity to become a part of this alliance.

On *May 22, 1920*, it was decided to be established "The International Radio Telegraph Company" on similar lines of RCA-GEC tie up and on *June 21, 1920*, The International Radio Telegraph Company" and the Westinghouse Electric and Manufacturing Company signed the agreement, however in this case, International Radio Telegraph Company was left powerless.

First time in *1922*, RCA started marketing crystal radio receiver Radiola models manufactured by General Electric (GE) in *August 1922*. The Radiola-I crystal radio set is shown in Fig. 6.35.

6.7.6 BBC Forms a Syndicate for Manufacture of Crystal Radio Sets

British Broadcasting Company formed by syndicate of manufacturers in *18 October 1922*, developed series of crystal radio sets with BBC logo for home users from *1922–1927* shown in Fig. 6.36. In *1925*, many circuits demonstrating the

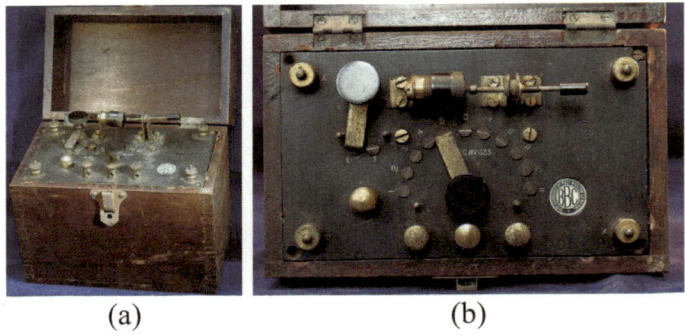

(a)   (b)

Fig. 6.36 Ediswan the makers of "Ediswan 1923B Crystal Set" made for BBC for home users, **a** Photograph of Ediswan Crystal set **b** Front panel of Ediswan Crystal set, Courtesy Allan Isaacs, Ediswan 1923B Crystal Set, http://www.radiomuseum.co.uk/radio2.html (Allen's virtual radio museum)

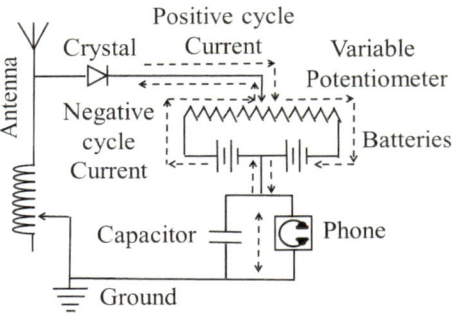

Fig. 6.37 Rectification properties of crystals

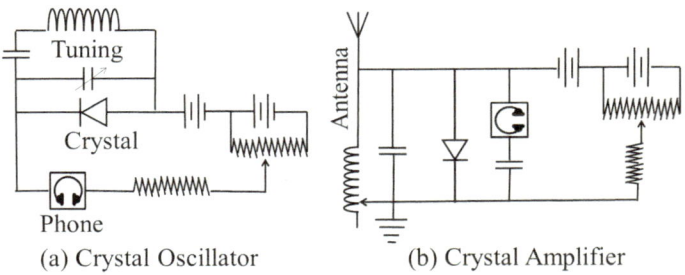

(a) Crystal Oscillator   (b) Crystal Amplifier

Fig. 6.38 Crystal oscillator and amplifier circuits

use of crystals such as design of crystal assembly, crystal oscillator, crystal high frequency amplifier, low frequency amplifier and radio receiver were being published [171] for the benefit of general public which is shown in Fig. 6.37. The crystal oscillator and amplifier circuits are shown in Fig. 6.38a and b.

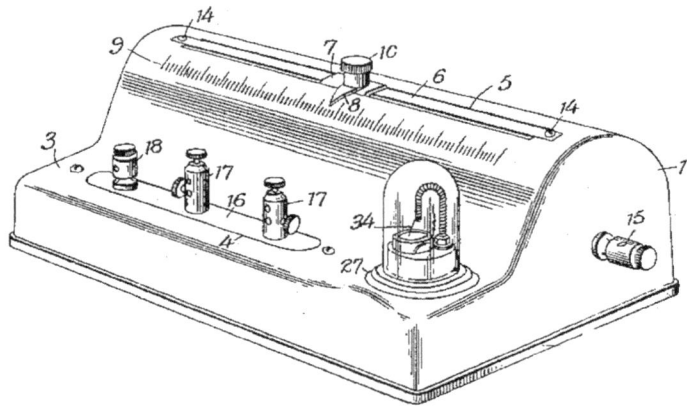

Fig. 6.39   Adams's Radio Apparatus based on crystal detector (United States Patent No. 1,748,435 dated February 25, 1930)

During *1920s*, many commercial companies started producing various types of crystal based radio receiver sets at cheaper rates [172] and hence market was flourished with these type of radio sets. These companies are toy manufacturer A. C. Gilbert; The Aerial Electric Co; Aerophone company; AGF; The Bethlehem Radio Corporation; Manhattan OPT in Rochester, N.Y; Intercity Marine W. E. Co. Inc., Seattle, Washington; Jubilee Manufacturing Company of Omaha, Nebraska; Lalley Light Corporation's Radio Apparatus Division in Detroit Michigan; Compania National de TS.H. in Madrid; Metropolitan Vickers Co.; Novelty Radio Manufacturing in St. Louis, Missouri; Foote Mineral Co., Philadelphia.; Rao Manufacturing Co. of Minneapolis, Radio Corporation of America; General Electric Company; Telefunken; Tiny tone Radio Company; Midget Radio Company and Pa-Kette Electric Company and Midway Company.

Accessories for crystal radios were either manufactured or supplied by Crain Bros. Radio Shoppe; Ajax Electrical Speciality Company; M. Carlton Dank & Co.; Palmer Electric and Manufacturing Co.; Ferro Manufacturing Company; Heard Company; Kodel Manufacturing Company; Parker Radio Company; Magna Kit Company; Furr Radio; Mohawk Battery and Radio Company; Rippner Brothers Manufacturing Company; Modern radio Labs; Philmore Manufacturing Company and Radio Marine Corporation of Am.

H Adams and Taylor Ottie B developed crystal radio set [173] and filed the patent for this radio set in on *March 12, 1926* and final US Patent No. 1,748,435 was issued on *February 25, 1930* which is shown in Fig. 6.39.

The Bethlehem radio corporation produced two types of crystal radio sets, one called "Crystal Dyne" and "World Crystal Set" similar to the crystal radio set developed by H Adams and Taylor Ottie B as shown in Fig. 6.39. Crystal radio sets developed H Adams and Taylor Ottie B and manufactured by Bethlehem Radio Corporation used crystal assembly similar to glass enclosed crystal assembly produced

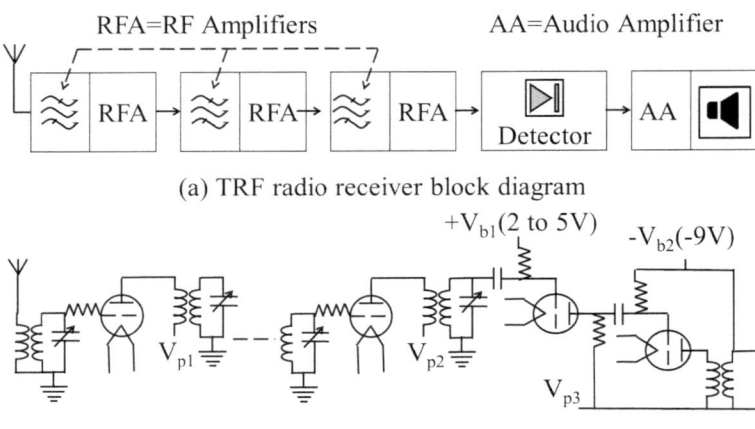

(a) TRF radio receiver block diagram

(b) TRF radio receiver Circuit diagram

Fig. 6.40 Tuned Radio Frequency Receiver (TRF)

by Fillmore Manufacturing Company, however it was not produced by Fillmore Manufacturing Company. The users of glass enclosed crystal assembly produced by Fillmore Manufacturing Company were Luxer, Peerless, Selective, Supertone and Top Notch of Insuline Corporation of Am.

6.7.7   Tuned Radio Frequency Receiver Developments

During *1920s*, the number of broadcasting stations working on different radio frequencies rose to new heights, various parameters like greater levels of performance, sensitivity and selectivity played a crucial role in competitiveness and user service demands. It was a time when vacuum tubes and improved tuning techniques were in demand.

Early vacuum tube-based radio frequency signal amplification relied on changing the filament current to increase the capacity of the amplifier, however, inter-electrode capacitance provided unwanted feedback that was responsible for creating the oscillations in the amplifier circuit that produced unwanted whistle sound in the received voice signal when one tried to set higher amplification in the individual amplification stage and hence to avoid this situation, multiple tuned amplifier stages at low amplification were used to gradually amplify the signal in subsequent stages this led to the developments of Tuned Radio Frequency (TRF) radio receivers shown in Fig. 6.40. This type of receiver was being tuned to a particular station frequency by ganged tuning of three sequential stages at once without adding any additional external frequency. The concept of Tuned Radio Frequency (TRF) was patented by Ernst Alexanderson in *1916*.

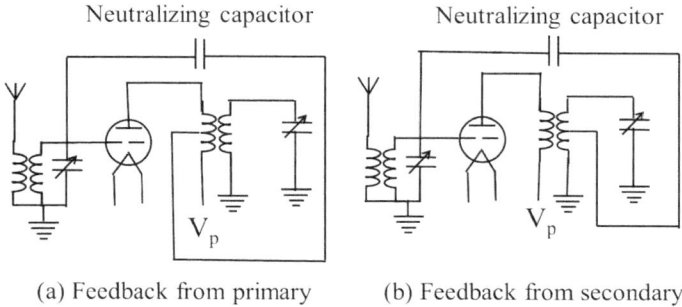

(a) Feedback from primary   (b) Feedback from secondary

Fig. 6.41 Tuned Neutrodyne Amplifier (TNA)

The unwanted inter-electrode parasitic capacitance between plate and grid was responsible for parasitic oscillations in early radio receivers. Professor Louis A. Hazeltine [174] at Stevens Institute of Technology, Hoboken, New York worked in the direction of reducing the effect of such parasitic capacitance and came with a circuit called "Neutrodyne". In *March 02, 1923*, Prof. Hazeltine gave a talk entitled "Tuned Radio Frequency Amplification with Neutralisation of Capacity Coupling". The circuit of Tuned Neutrodyne Amplifier(TNA) is shown in Fig. 6.41.

At this point broadcast industry was facing the problem of reception of extremely weak spark signals ranging from 500 kHz to 3 MHz which required elaborate adjustment and could not be done using minimum adjustments. Due to the work of Round and Latour, the signals of these frequencies up to 2 MHz can now be amplified using vacuum tubes and transformers of low capacity [148] and these limitations were responsible for limiting the use of superheterodyne techniques. Superheterodyne principle invented in *1918*, and its application to shortwaves with full details appeared in *1919*. Superheterodyne techniques had always been superior in performance in terms of selectivity and sensitivity. The further improvements were possible due to the advent of their improved circuit forms, started finding their way into radio receiver applications.

6.7.8 AT&T Joins Cross Licensing Group

After the establishment of Radio Corporation of America in *1919* and signing cross licensing agreements with GE and AT&T in *June 1920*, picked up the commercially manufactured radio market, Westinghouse could sense the situation that the commercially manufactured radio market is slipping from their hands, turned their attention to broadcasting business to create a respect for their own manufactured radio receiver sets in the market and also purchased regeneration and superheterodyne patents of Armstrong in *October 1920* to consolidate their position in the radio receiver market.

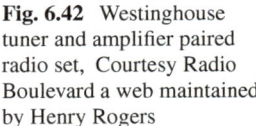

Fig. 6.42 Westinghouse
tuner and amplifier paired
radio set, Courtesy Radio
Boulevard a web maintained
by Henry Rogers

6.7.9   Developments at Westinghouse and Joining of Cross Licensing Group

At Westinghouse, a pair of RA type high grade short wave (180–700 m) regenerative tuner and detector amplifier (DA) designed by Frank Conrad and Donald Little went in production late *1920* [169]. The apparatus consisting of this pair provided the most efficient radio set for telegraph and telephone reception for armature and ship wavelengths and is shown in Fig. 6.42. This paired radio set using advanced techniques had been a great success that provided a great opportunity to Westinghouse to negotiate and join cross licensing group (RCA, GE and AT&T).

In *1921*, Westinghouse also joined this cross licensing patent group [175] wherein GE and Westinghouse had exclusive rights for manufacturing radio receiver sets, AT&T had exclusive rights to manufacture, lease and sell transmitters and RCA had exclusive rights to sell radio receiver sets.

In mid 1921, Westinghouse enclosed RA and DA pair (RA-DA) in a single cabinet and called this model as RC. Westinghouse also marketed Antenna Tuner (RT) and an RF amplifier (AR) for matching RA-DA pair. The RD-DA pair uses in all three vacuum tube valves one UV-200 soft detector and two UV-201 hard amplifiers[176] shown in Fig. 6.43a, b. From December 1921, RCA being a member of cross licensing group began to advertise the radio equipment manufactured by Westinghouse even though Westinghouse continued to advertise RA and DA till March 1922, and during the next two and half years, large number of units of RA and DA were produced.

In 1921, Westinghouse introduced Aeriola Junior (Fig. 6.44a) crystal set receiver while in later part of the year that is on December 1921, a wooden case Aeriola Senior (Fig. 6.44b, d, e) tuned radio frequency receiver regeneration principle was introduced by Westinghouse, however, since WD-11 was delivered first on January 1922, the radio receiver that integrates WD-11 (Fig. 6.43c) was ready only on January 1922,

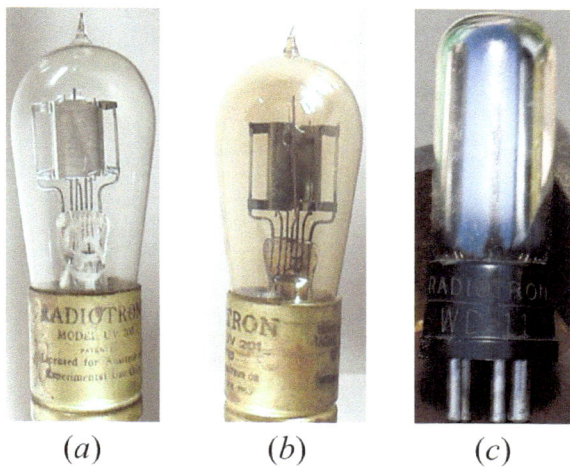

(a) (b) (c)

Fig. 6.43 UV-200 vacuum tube and UV-201 Valves used in RA/DA-Pair and WD-11 valve used in Aeriola and Radiola receivers **a** UV-200 **b** UV-201 Courtesy: Bruce Morgenstern and the Radio Museum **c** WD-11 Vacuum tube, Courtesy: James Ollinger web

RCA started its marketing only after January 1922. The Aeriola Senior, was also supplied with additional two stages of audio amplifiers with sufficient gain required for driving the horn speaker. Both Aeriola Junior and Radiola Senior were single tube receivers worked on the regeneration principle and used a single WD-11 tube. In the same year, that is, in 1922, Aeriola Grand (Fig. 6.44c) was also introduced.

6.7.10 Hoover's Radio Conferences for Streamlining the Use of Radio Frequency Waves

Herbert Clark Hoover (President of US from *March 4, 1929* to *March 4, 1933*), engineer, businessman and politician, who served as US secretary of commerce from *1920* to *1928* during US presidents Warren G. Harding and Calvin Coolidge felt that the rampant and directionless growth wireless radio systems at that time need to be checked. It was observed that in the previous year of *1922*, the number of radio receivers increased from 50,000 to 6,00,000 and hence Herbert Clark Hoover, with the intension of strengthening the relations between industry and government, arranged first radio conference called "Hoovers's Radio Conference" in *February 27 1922*. He was a believer of the fact that the regulation of broadcasting is a must to avoid interference amongst various radio stations. He was with the opinion that "broadcasting material must be limited to news, to education, and to entertainment, and the communication of such commercial matters as are of importance to large groups, of the community at the same time". He carefully planned successive four annual confer-

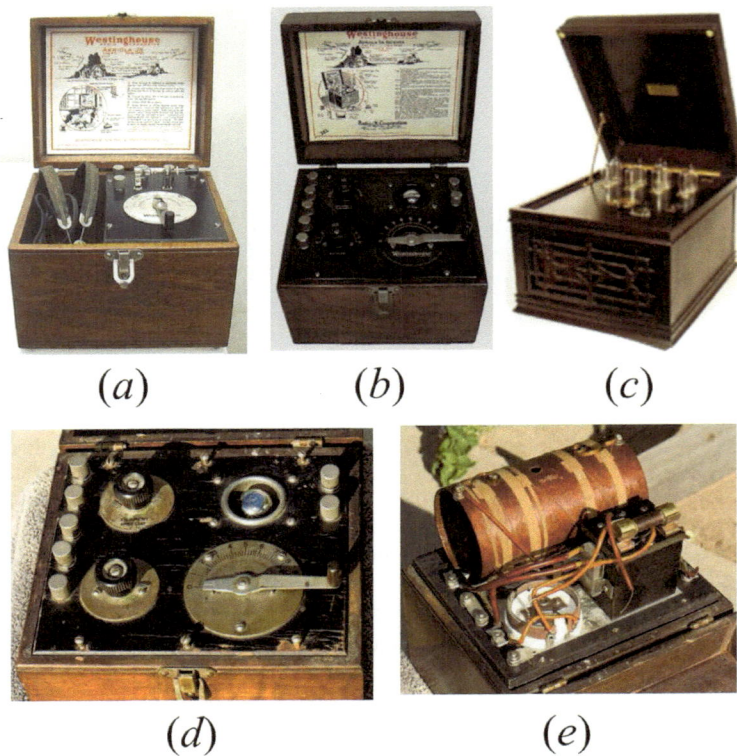

(a) (b) (c)

(d) (e)

Fig. 6.44 Aeriola Junior, Credit: the RadiolaGuy **b** Aeriola Senior Credit: Vintage Radio Web maintained by John M Koster **c** Aeriola Grand, Courtesy: TechnoGallerie.com **d** Aeriola Senior Front Panel View Courtesy: James Ollinger web and **e** Aeriola Senior internal circuit Courtesy: James Ollinger web

ences from *1922* to *1925* that lead to Federal Radio Act in an international conference of *1927*, called "Federal Radio Act of *1927*" which became Federal Communication act of *1934* that deals with regulation of frequency spectrum.

The first two National Radio Conferences on *February 27, 1922* and *March 20–24, 1923*, resolved [177, 178] that "radio communication as a public utility and should be regulated and controlled by Federal Government in public interest" and Radio equipment that effectively reduces the interference should be promoted and should be made freely available to the public. It had specified different wavelengths/frequencies for different uses as shown in Table 6.5.

The later two National Radio Conferences on *October 6–10, 1924* and *November 9–11, 1925*, [179, 180] to focus on various issues related to **Telegraphic Radio:** Expansion of international communication by radiotelegraph, Clearing of broadcasting band of coded signals, **Telephonic Radio**, **Service Area**, **Scientific Investigations**, **Industry Problems:**interconnections, advertising, Removal of Sta-

Table 6.5 Wave frequency allocations for various uses

Type of use	Wavelength (Meters)	Frequency (kHz)
Transoceanic Telephone Experiments	6000–5000	50–60
Fixed Service Telephony	3300–2850	90.9–105.2
Mobile Radio Telephony	2650–2500	113.2–120.0
Govt. Broadcast	2050–1850	146–162
Fixed station radio telephony	1650–1550	181.8–193.5
Aircraft radio telephony and telegraphy (E)	1550–1500	193.5–200
Government and public broadcasting	1500–1050	200–285.7
Radio beacons (E)	1050–950	285.7–316
Aircraft radio telephony and telegraphy (E)	950–850	316–353
Radio compass service (E)	850–750	353–400
Govt. Pub Broadcast 200 mile or more from seacoast (E)	750–700	400–428
Govt. Pub Broadcast 400 mile or more from seacoast (E)	700–650	428–462
Marine radio telephony	750–650	400–462
Aircraft radio telephony and telegraphy (E)	525–500	572–600
Government and public broadcasting, exclusive (E)	495–485	606–618
Private and toll broadcasting	485–285	618–1052
Restricted special amateur radio telegraphy	310	968
City and state public safety broadcasting (E)	285–275	1052–1091
Technical and training schools & Amateur	275–200	1091–1500
Amateur telegraphy and telephony (E)	275–150	1091–2000
Private and toil broadcasting (E)	150–100	2000–3000
Reserved	100	3000

tions From Congested Areas By Remote Control, **Problems For Solution By Cooperation With The Government:** overcrowding in air, actions to be taken. It had specified different continuous waves (CW) and "Interrupted Continuous Wave" or ICW or on-off keying (OOK) for various types of uses are shown in Table 6.6. Based on the recommendation of the third radio conference, broadcasting below 550 kHz

Table 6.6 Wave frequency allocations for various uses based on transmission types

Type of use	Transmission type	Wavelength (Meter)	Resp. Freq. (kHz)
Government	CW	2399	125
	CW, ICW	1934, 1713, 1224, 1090, 674	155, 175, 245, 275, 445
Government only	CW, ICW	3156-2499, 1578-1304, 952	95–120, 190–230, 315
Marine Only	CW, ICW, Sprk	731, 706, 660	410, 425, 454
	CW, ICW	874	343
	Phone	1276–1052	235–285
Marine & aircraft	CW, ICW	2499–1960	120–153
P2P Marine & arircraft only	CW, ICW	1960–1817	153–165
P2P Marine Only	CW, ICW	1817–1578	165–190
Univ & College Expt. only	CW, ICW	1304–1276	230–235
Beacons Only	CW, ICW	1000	300
Radio Compass Only	CW, ICW	800	375
Distress Calling and messages only	CW, ICW, Sprk, Phone	600	500
Aircraft and fixed safety of life stations	CW, ICW, Phone	600–545	500–550
Broadcast Only	Phone	545–200	550–1500
Amateur Only	CW, ICW, Phone	200–150	1500–2000

was eliminated and the broadcasting range of 550–1500 kHz was divided into three classes of radio broadcasting stations called Class-I, Class-II and Class-III stations. The Class-I Stations operating in the range of 550 to 1070 kHz (63 Channels), Class-II Stations operating in the range of 1090 to 1400 kHz (32 Channels) and Class-III Stations operating in the range of 1420 to 1460 kHz (5 Channels).

The wave frequency allocations for various uses with respect to the number of transmission types are shown in Table 6.7.

The fourth radio conference Secretary of Commerce was practically stripped of all the powers of the policy of issuing licenses to all and favoured for the industry to solve their problems with private partnership rather than depending on government initiatives, however conference was unanimous about Government's role in reducing the number of radio broadcasting stations.

Table 6.7 Wave frequency allocations for various uses with no transmission types

Type of use	Wavelength (Meter)	Respective Freq. (kHz)
Marine & Costal Only	1052–600	285–500
P2P	150–133	2000–2250
Aircraft only	133–130	2250–2300
Mobile & Govt. Mobile	130–109	2300–2750
Relay broadcasting only	109–105, 66.3–60.0, 54.5–52.6, 33.1–30.0, 27.3–26.3	2750–2850, 4525–5000, 5500–5700, 9050–10000, 11000–11400
Pub Toll, Govt.Mob., P2P,P2P Elect.Supp, Mult. Addr. Msg. Press only	105-85.7	2850–3500
Amateur, Army Mob., Naval Aircraft, Navy Vessel Aircrsft only	85.7–75	3500–4000
Mob., Pub Toll, Govt. P2P, P2P Pub. Util.	75–66.3, 37.5–33.1	4000–4525, 8000–9050
Pub Toll only	60–54.5	5000–5500
P2P only	52.6–42.8	5700–7000
Amateur and Army mob. only	42.8–37.5	7000–8000
Pub Toll only	30–27.3	10000–11000
Pub Serv, Mob., P2P Govt.	26.3–21.4	11400–14000
Amateur only	21.4–18.7	14000–18000
Experimental	16.6–5.35, 4.69–0.7496	18100–56000, 64000–400000
Pub. Toll, Mob., P2P Govt.	18.7–16.6	16000–18100
Amateur	5.35–4.69, 0.7496–0.7477	56000–6400, 400000–401000

6.7.11 Valve Based Radio Receiver Architectures

The three types of radio receiver architectures called Autodyne, Tuned radio frequency and Superheterodyne receivers have emerged, which are described in the following paragraphs.

6.7.11.1 Autodyne Receiver Architecture

Autodyne is a form of a circuit that is tuned slightly off the centre of in relation to a frequency of the signal of interest. This circuit acts as a combined local oscillator, amplifier and demodulating carrier wave circuit. In this case antenna is connected to regenerative through an untuned RF stage and band pass filter. Block diagram of Autodyne Receiver is shown in Fig. 6.45.

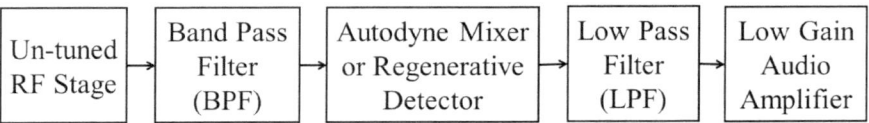

Fig. 6.45 Autodyne radio receiver architecture

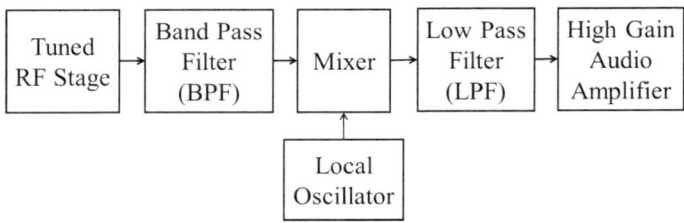

Fig. 6.46 Homodyne radio receiver architecture

Autodyne receivers are critical and sensitive because of its autodyne mixing stage and background hissing noise can not be brought down to the desired level.

6.7.11.2   Direct Conversion Radio Receiver Architecture

These receivers are also called as homodyne or synchrodyne or zero-IF radio receivers wherein it employs synchronous detection principle local oscillator is tuned to incoming RF frequency. Block diagram of Homodyne Receiver is shown in Fig. 6.46.

6.7.11.3   Superheterodyne Radio Receiver Architecture

The early radio circuits were directly tuned to a given incoming frequency while in heterodyning, an additional oscillator was required to tune to various frequencies that created a beat frequency or intermediate frequency that is amplified, detected to extract audio signals which are further amplified and produced in the form of sound.

After elaborate experimentation eight tube superheterodyne radio receiver set was constructed, consisting of heterodyne oscillator, three stages of intermediate frequency (IF) amplifiers, rectifier or detector and two stage audio amplifier. IF stages ere coupled using air core transformers. The block diagram is shown in Fig. 6.47.

6.7.12   Radiola Series of Radio Receivers

The term called "Radiola" series of receivers was proposed by Alfred N. Goldsmith working at RCA research in New York. The production was started from 1921 by the league of companies for which RCA was marketing the radio receivers. From

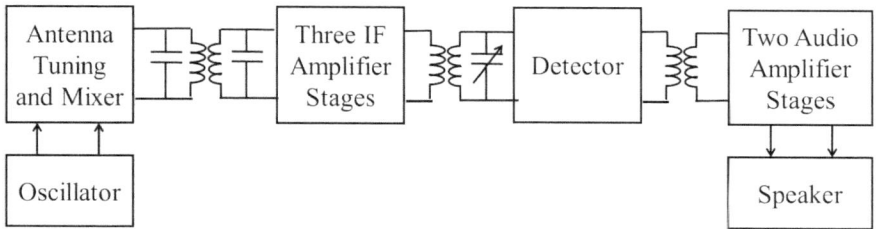

Fig. 6.47 Building blocks of superheterodyne receiver

May 1923, RCA started marketing of vacuum tube- based Radiola models [181] those were mass produced by General Electric (GE), Westinghouse and RCA-Victor company for marketing by RCA and remained in market from *1921* to *1931*.

6.7.12.1 Radiola RT, RC, AR, RS, SR and AC Models

Radiola-RT antenna coupler that was used as Pre-Tuner in front of radio amplifier (RA) which was further connected to detector and amplifier (DA) to make the configuration RT+(RA+DA)identical to RT+RC. The three stage radio amplifier using three valves was designed as Radiola-AR while earlier Radiola-RA used one UV-200 valve (radiotran).

Radiola-RS receiver had a provision for connecting long wave and shortwave antennas and implemented two valves (WD-11), the phone outlet is provided for connecting headphone while Radiola-SR receiver also has provision for connecting long wave and shortwave antennas and implemented using one valve (WD-11) and phone outlet for connecting headphone. Both Radiola-RS (1923) and Radiola-SR can be further connected to Radiola-AC two stage audio frequency amplifier. Radiola-AC, Radiola-RS, and Radiola-RT were produced in an interleaved manner.

6.7.12.2 Radiola-1: Valve-Crystal (VC) Models

Radiola-1 was typically a valve and crystal or solid state detector radio receiver model that used one audio frequency amplifier stage and only catered long wave and medium wave only as shown in Fig. 6.35.

6.7.12.3 Radiola-II Models-TRF with Regeneration

The first of its own type self contained portable radio set was Radiola-II Model AR800 that included antenna, two UV-199 Tubes, headset and batteries. The photographs of Radiola-II radio set are shown in Fig. 6.48. Radiola-II sets were based on the principle of tuned radio frequency(TRF) with regeneration. Radiola-II Grand model

(a)   (b)   (c)

Fig. 6.48 External and internal views of Radiola-II receiver sets **a** AR 800 Model front panel view, Credit: John D. Jenkins, Sparkmuseum **b** Radiola-II, AR 800 Model internal view Credit: IARCHS Radio Collector Club and **c** Westinghouse for RCA Radiola Grand—Model RG, Credit: Radio Boulevard, a web maintained by Henry Rogers

was a four tube implementation catering to tuning of long wave and shortwave stations (Fig. 6.48c).

6.7.12.4   Radiola-III Models:2-Valve TRF with Regeneration

During *January 16, 1923*, Research Department of Radio Corporation of America (RCA), provided a document called "Specifications for Broadcast Receivers for 1924" provided initial thoughts on the 1923–1924 models to be manufactured by to its group of manufacturers. The specifications Radiolas III, IIIA, Regenoflex, X, Superheterodyne and Super VIII were announced, planned and possibly developed but never manufactured.

Radiola-III (also called Type RI) with a tuning range from 250 m (1199.16 KHz) to 550 m (856.549 KHz) were existed during *1924*. These were vacuum tube radio receivers based high quality regenerative radio receivers those used two WD-11 tubes having cleaner amplification and adaptable to antennas of various sizes. The photograph of Radiola-III radio set is shown in Fig. 6.49.

There were three types of Radiola-III models called Type RI, AR-805-Radiola-III Type RL (Fig. 6.49a) and Radiola-IIIA integrates both AR-806-Radiola-IIIA Type RF & Radiola Balanced Amplifier together using four WD-11 vacuum tube valves (Fig. 6.49b and c).

6.7.12.5   Radiola-IV to VII Models

Radiola IV (Fig. 6.50a, b) [182], although specified in *July 25, 1922* was introduced in *December 1922*, and 10,400 units were made. It was one of few early radio receivers enclosed in a fine wooden cabinet with two wooden doors made from the wood of Mahogany tree. The receiver was based on Tuned Radio Frequency (TRF) with regeneration principle and used three valve tubes of type UV-199, one tube is used

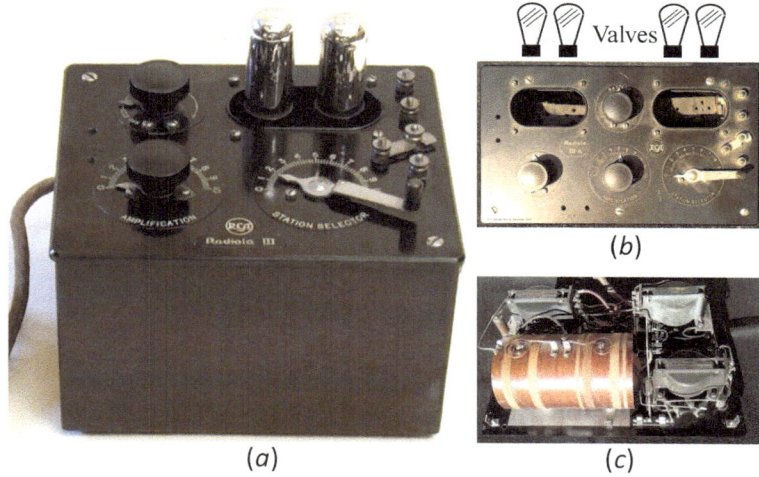

Fig. 6.49 Photograph of **a** Radiola-III Receiver set, Courtesy: Phil's old Radios:RCA Radiola Model III Radio (1924), antiqueradio.org **b** Radiola IIIA receiver set and **c** Inferior of Radiola IIIA receiver set manufactured by RCA-Victor Company Inc., Courtesy: Š Murray Greenman, My HF Receiver Collection, QSL.NET: Connecting Hams Around The World, May 24, 2019

for detection purpose and other tubes are used for audio frequency stages. It has two antenna jacks, one for the wavelength range of 200–400 m and the other for a range of 350–600 m.

From 1922, RCA marketed RCA Radiola V (Fig. 6.50c, d) Type AR685A (1922), GE AR 1300 (1922), GE AA-1400 (1922) and AR885 (Jan 1924). RCA Radiola AR885, when used with average an antenna, it can be tuned to wavelengths from 180 m to 700 m but when a long wave coil system (Model UL-1340) is used, a set can be used to receive wavelengths from 650 to 1150 and 1450 to 2800 m wavelengths, it also used regenerative tuning using one UV-200 or UV-201 vacuum tube detector and two stages of audio signal amplification using two UV-201 valves.

Radiola VI Model AR-895 of 1923 (Fig. 6.50e) implemented using three stages of RF amplification using UV-200 or UV-201A, Detector using one UV-200 or UV-201A vacuum tube and two stages of AF amplification using UV-200 or UV-201A, vacuum tube. In all 6 vacuum tubes are used to tune the wavelength from 300 ms (10,000 KHz) to 5000 m (60 KHz) using variable air condenser varying from 0.0007 to 0.001 microfarad. The two ranges of wavelengths from 200 to 500 m and 500 to 5000 m selected using wavelength selector switch and its outer look was like Radiola V.

Radiola-VII models (Fig. 6.50f) were introduced from 1921 and its variants were Radiola VIIA and Radiola VIIB. RCA-Victor Company Inc. produced RCA Radiola VII MODEL AR-905(September 1923), Radiola VIIA September 1923 (Fig. 6.51a) possibly never sold and Radiola VIIB Model AR907 1924 (Fig. 6.51b). The Radiola VII was implemented as a self contained unit using five tubes of type UV-199 and

Fig. 6.50 Photograph of **a** Exterior View of Radiola-IV, Credit: the RadiolaGuy **b** Exterior view of Radiola-IV with control knobs exposed Credit: John D. Jenkins, Sparkmuseum **c** Exterior View of Radiola V, Credit : Steve Erenberg, radio-guy **d** Interior View of Radiola V, Credit: Antique-Radios.net: Information site for antique radio collectors **e** Exterior View of Radiola VI Credit: John D. Jenkins, Sparkmuseum and **f** Exterior View of Radiola VII receiver sets marketed by RCA, Credit: John D. Jenkins, Sparkmuseum

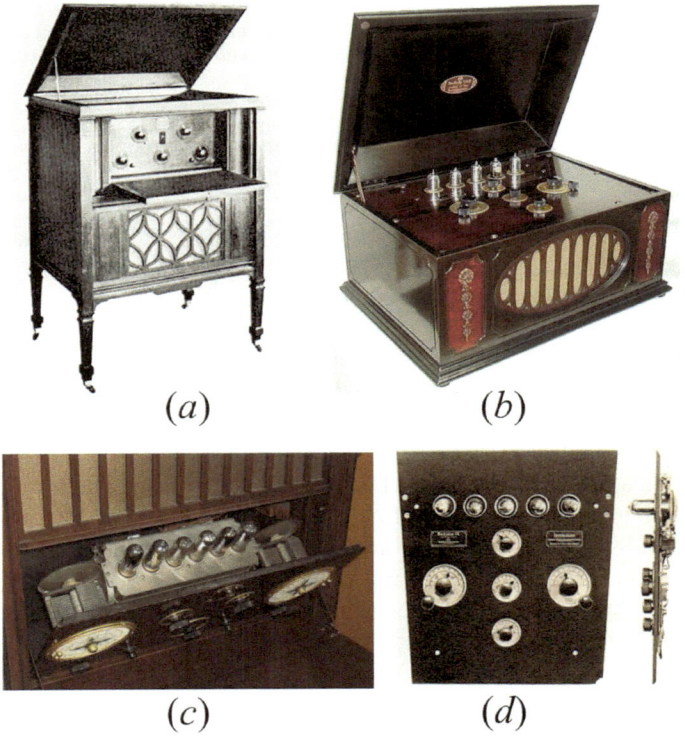

Fig. 6.51 Photographs of **a** Radiola-VIIA and **b** Radiola VIIB **c** Radiola VIII **d** Radiola IX, *Credit* All the figures are by the courtesy of the RadiolaGuy

consisted of the highly selective tuner and five tube detector amplifier and ready to connect loop antenna. The Radiola VIIB is completely enclosed in a wooden cabinet that contains a complete receiver including both batteries and loudspeaker and covers a broadcasting band of 220 to 550 m ensuring uniform sensitivity throughout the entire band.

The Radiola Model VIII is shown in Fig. 6.51c while Radiola-IX Model (Fig. 6.51d) (RCA Catalog of 1924) is also implemented similar to Radiola-VIIB using five UV-199 Radiotrons. It also uses a highly selective two circuit tuner, detector and radio audio amplifier and tunes to the wavelengths from 220 to 550 m.

6.7.12.6 Radiola X and Radio Regenoflex Models

The both Radio X (Fig. 6.52a) and Radiola Regenoflex (Fig. 6.52b, c) models have similar specifications and work on the principles of regeneration. These model use four WD-11 vacuum tubes (Radiotrons) with built in batteries and speaker that had good sound and tone quality mounted in an attractive mahogany cabinet. The radio operates on wavelengths from 220 to 550 m or frequencies from 1400 to 540 KHz. The common characterises of these receivers are higher selectivity, sensitivity to distant stations, Audio frequency amplification using special alloy transformers, work on regeneration principles, free from radiation, self contained cabinets, The main differences are that in Regenoflex model uses one Radiotron tube to amplify both Radio and Audio frequencies while for audio amplification balanced amplifier is used while Radiola X model uses separate radio and audio amplifications and minor mechanical changes.

6.7.13 Spread of Vacuum Tube Based Superheterodyne Radio Receivers

One of the first prototype superheterodyne radio receivers built by inventor Edwin Armstrong. The superheterodyne, the circuit used in virtually all modern radios, was invented by Armstrong in *June 05, 1918* while he worked in a US Army Signal Corps laboratory in Paris during World War 1. This is one of the receivers that was constructed at that laboratory is described in an article of Radio Amateur News magazine of *February 1920*. The superheterodyne receiver was first evolved by Edwin Howard Armstrong in 1918 and was introduced in the market in 1920.

Earlier superheterodyne receivers used a large number of vacuum tubes and hence in *June 05, 1918*, Armstrong and his team tried to make an oscillator tube that can be used for two types of jobs in France but were not very successful due tuning problems. Harry Houck solved this problem by mixing the received signal with the oscillator signal of the second harmonic.

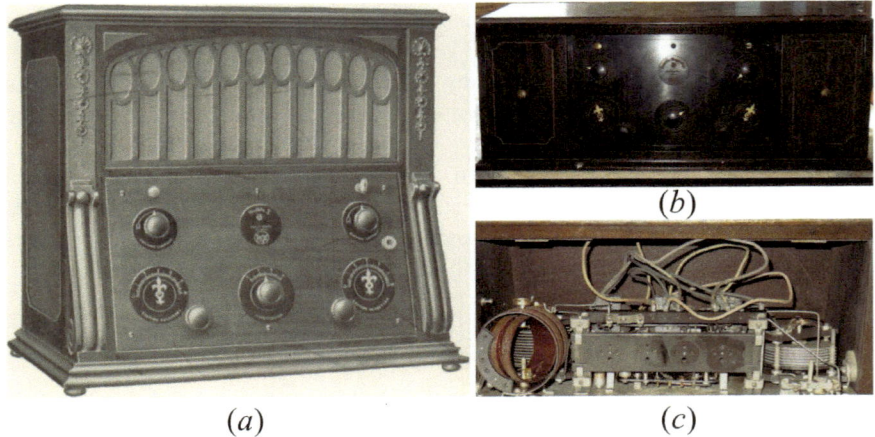

(b)

(a) (c)

Fig. 6.52 Photographs of **a** Radiola-X, Courtesy: By permission of the California Historical Radio Society and its program, the Society of Wireless Pioneers. **b** The front panel view of Radiola Regenoflex and **c** Inside view of Radiola Regenoflex and Courtesy: Antique Radios: The Collectors Resource

6.7.13.1   Radiola VIII Superheterodyne Radio Receivers Models

RCA introduced four types of superheterodyne receivers called Radiola Super VIII (Type AR-810) in February 1924, Semi-portable (Type AR-812) in February 1924, Radiola 24 and Radiola 26 in the form of 6 tube configurations. Radiola Super VIII (Type AR-810) was one of the first Superheterodyne receivers and RCA's first console radio

The first superheterodyne radio receiver in Radiola series called "Radiola Superheterodyne AR-812" was brought by two engineers Edwin H. Armstrong and Harry Houck in _March 04, 1924_ for marketing by RCA. In which a number of tubes were brought down from eight to six from the first superheterodyne receiver model developed by Edwin H. Armstrong during _1917_. The photograph of Radiola Superheterodyne AR-812 is shown in Fig. 6.53.

The circuit diagram of Radiola "Superheterodyne" AR-812 is shown in Fig. 6.54.

6.7.14   Networking of Radio Broadcasting Stations

There were two options before broadcasting companies to increase their area of coverage was to increase the power of transmission or to interconnect their other broadcasting radio stations for retransmission. Some companies opted for a natural choice of increasing the power of radio stations while the world's largest company AT&T adopted the second approach.

Fig. 6.53 RCA semi-portable Superheterodyne Receiver Radiola VIII (AR-812), Courtesy Radio Boulevard: Western Historic Radio Museum maintained by Henry Rogers

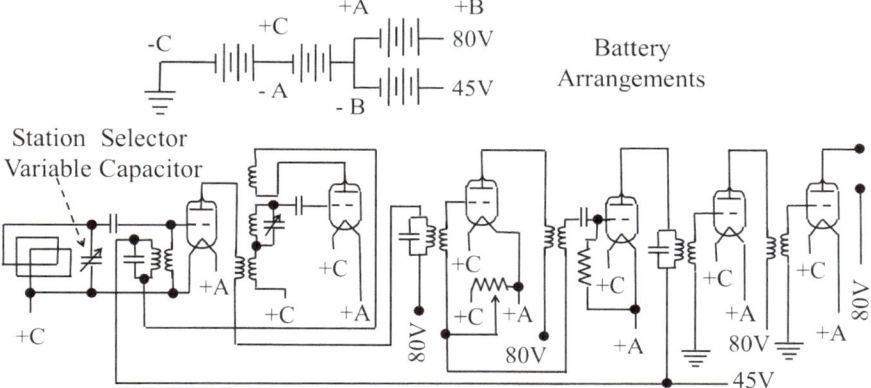

Fig. 6.54 Circuit diagram of RCA Radiola Superheterodyne AR-812

6.7.14.1 GE's Experiments of Retransmission

Earlier experiments were to confirm that how long radio reception can be received. In early *April 1919*, General Electric put its efforts to connect US Navy's station with call sign NFF at New Brunswick, New Jersey with low power vacuum tube based transmitter installed aboard U.S.S. George Washington sailing in the Atlantic ocean which also had been installed with Harold H. Beverage's "Duplex Radio-phone Receiver. The low power signal transmitted by U.S.S. George Washington was received by NFF station were retransmitted by high power transmitter of NFF which were loudly heard back at U.S.S. George Washington was an example of hooking up of two radio stations using 167 kHz longwave.

6.7.14.2   US Navy's Joint Broadcast

The US Navy's radio station with call sign NAA at Arlington started vacuum tube technologies in early *1920*, and on *May 30, 1922* NAA in conjunction with another Navy station with call sign NOF Anacostia, D.C, jointly broadcasted a dedication ceremony of Lincoln Memorial. Both the radio stations NAA and NOF were connected by the telephone line provided by the local telephone company.

6.7.14.3   AT&T Develops Network of Radio Stations

AT&T was already mindful in putting its efforts in the direction of building the network radio stations since early *1923*, and started connecting radio stations using telephone lines for combined transmission to increase the physical area of coverage of radio transmission to target country wide transmission program. *January 4, 1923*, it started an experiment to connect two radio stations WEAF in New York and WNAC (later become WRKO) in Boston, Massachusetts by temporary lines. Both stations successfully transmitted a combined program for three hours. A radio station WMAF of Round Hills Radio Corporation in South Dartmouth, Massachusetts licensed on September *1922*, owned by one of the wealthiest personality H. R. Green paid money to AT&T on *July 1, 1923* to establish first fixed link between WEAF and WMAF for the purpose of simultaneous broadcasting of WEAF programmes. From the summer of *1923*, it rebroadcasted the night programmes of WEAF station in New York City. By early *1926*, WEAF chain was spread to 19 cities and AT&T becomes important broadcast networking company.

6.7.14.4   Contribution of "Radio Group" for Networking of Radio Stations:

"Radio Group" consisting of General Electric, Westinghouse, and their jointly-owned subsidiary, the Radio Corporation of America was headed by its General Manager, David Sarnoff was maintaining a station WJZ with its status as "Super Power" due to its transmitting power of 50 kW. The higher power did not help much to increase the physical area coverage and decided to AT&T way of linking various radio stations but AT&T expressed its inability to provide its long telephone line infrastructures for interconnection of RCA radio stations and hence solution of leased lines was adopted in *October 7, 1922*, that worked for shorter distances but results for longer distances were not successful due to non-availability of acoustic experts and expertise installation of telephone lines resulting in humming problems.

   The alternatives to telephone lines were explored by taking the motivation and clues from amateurist's work on short wave communication and its suitability for longer distance coverage due to reflection of waves from ionosphere for connection between broadcast radio stations. The short wave solutions seemed to be the most viable alternatives for interconnecting scattered radio stations.

W. W. Rodgers of Westinghouse reviewed the work of company about the connection of KDKA in East Pittsburgh, Pennsylvania with other two radio stations called KDPM in Cleveland, Ohio and WBZ in Boston, Massachusetts that explored the role of short waves in interconnecting the radio stations and concluded that National Broadcasting System (NBS) could be a reality within very few years.

6.7.14.5 Establishment of National Broadcasting Company Inc.

After world war I, RCA was formed by a conglomerate of American companies keep upper hand over radio broadcasting business and became a leading manufacturer of radio receivers in the world. In *1923*, David Sarnoff wrote a memorandum for founding of National Broadcasting Company, Inc to put before the board of Directors of RCA, with intention of targeting broadcasting business in large scale that will play its role as a world leader in broadcasting business that will also cover future TV broadcasting business. AT&T, primarily in the business of telephone systems, its embryonic WEAF radio station catering to the networking of broadcasting station decided to sell this station to RCA at the cost of $1 millon. RCA was also maintaining its own station WJZ catering to networking since *June 17, 1922*, with these flagship stations WEAF and WJZ in its control, RCA established a new company called National Broadcasting Company Inc. on *15 November 1926*. Both WEAF and WJZ networks of NBC started operating simultaneously side by side.

NBC's first broadcast was held for four and half hours covering music, comedy and talent of a day aired on *November 15, 1926* from New York using a network of 25 stations.

On *January 1, 1927*, NBC formally divided these networks in to two networks called "Red Network" and "Blue Network". The network associated with the stations hooked up with WEAF was red network while the network associated with WJZ was a blue network. Red networks catered to commercially sponsored broadcasting while blue network catered to non-sponsored activities of social importance covering news, cultural programmes and history.

The new year *1927* broadcast of NBC covered an annual Rose Bowl football game event in California. Another important broadcast was held on *June 11, 1927* after the success of historic transatlantic flight by Charles A. Lindbergh. In *1928*, NBC started covering national political events, Republican and Democratic national conventions.

6.7.14.6 Some Efforts in the Direction of Establishment of High Power Broadcasting Stations

The General Electric's radio station 2XAG (Later on named WGY) operating on frequency 810 kHz at Schenectady NY become first radio station in US to broadcast using 50 kW transmitter in *1925*, was upgraded to 100kW on *August 4, 1927* and was given a restricted permission to operate between 1.00 pm to 2.00 pm for 30 days.

From *July 1921*, radio manufacturer Powel Crosley Jr. started test on his 20W transmitter from his college hill home station with call sign 8CR started broadcasting "Song of India". On *March 22, 1922*, Crosley Broadcasting Corporation got the licence of 50W transmitter using call sign WLW, increased to 500 W in September *1922*; 1kW in *May 1924*; It became first 5kW broadcasting station *January 1925*; 50 kW in *October 4, 1928*; and trails of 500kW starts on *October 4, 1928* to become fully operational on *May 2, 1934*. The 500 kW operation of WLW was finally ended on *March 1, 1939*.

6.7.14.7   Establishment of Colombia Broadcasting System (CBS)

A New York city talent Arthur Judson created "United Independent Broadcasters" network in *January 27, 1927*, needed additional investors and this need was satisfied by "Columbia Phonograph Company" in *April 1927* that resulted in the renaming of "United Independent Broadcasters" to "Columbia Phonographic Broadcasting System" went in air on *September 18, 1927* with Howard L. Barlow Orchestra from a flagship radio station called WOR (**W**orld **O**f **R**adio) which had a network of its fifteen affiliates.

Historically, WOR was owned by Bamberger who started it on *February 22, 1922*, using a 500-watt transmitter working on 360 m (833 kHz) wave and went in air from Bamberger's department store. The call sign WOR earlier used by the ship SS California owned by Orient Lines was reissued to Bamberger. Initially, WOR operated for limited hours due to time sharing with stations WDT and WJY operating on the same wavelength of 360 m (833 kHz), WOR changed its frequency to 740 kHz in *June 1923* and started time sharing with WJY till *July 1926* when WJY signed off for good fortune of WOR to get full benefit of frequency.

In December *1924* WOR was successful in acquiring a studio in Manhattan gave the benefits of services of the stars in New York. In *1926*, it was further located back to its original York City studio at Chickering Hall at 27 West 57th Street to 1440 Broadway, near Times Square. With these prospects on the way, WOR moved to 710 kHz on *June 17, 1927*. Then WOR becomes the first charter member of CBS Radio Network. On December *1928* CBS acquires WABC and later on in *1934*, CBS partnered with WGN, Chicago and WLW, Cincinnati radio stations to form an alliance to form a mutual broadcasting system. On *March 4, 1935*, President Roosevelt inaugurates WOR's 50 kW directional transmitter.

6.7.14.8   BBC Becomes Public Corporation in 1927

John Reith was a founder and first manager of British Broadcasting Company (BBC) when it was setup in *1922*. On *Janurary 01, 1927*, British Broadcasting Company became a public corporation and John Reith became the first Director General of British Broadcasting Corporation (BBC). BBC started the expansion of its network of regional transmitters. The BBC's national transmitter 5XX working on 1600 m

(187.37 kHz) and broadcasting since *1925* was then used for networking of various local transmitters and was joined by Empire Service Transmitters (forerunner of present day BBC world service).

6.7.15   The Air Commerce Act of 1926

In the United Sates of America, The Air Commerce Act [183] was established and came in existence in 1926 called "The Air Commerce Act of 1926" and became a law in May 1926, as an effect of a serious study initiated by the aircraft industry that forced the involvement of legal bodies like American Bar Association and other bodies to work on Law of Aeronautics. This facilitated to work on aircraft altitude separation, develop and maintain airways and navigational aids.

6.7.16   Radio Act of 1927

The Radio Act of *1912* gives regulatory power to the United States Department of Commerce. The Commerce Secretary, Herbert Hoover played a crucial role in shaping the future of radio communication. However, the powers of commerce secretary were limited by federal court decisions and had no powers to deny the licenses created a chaotic situation leading to too many transmitting stations. In order to rectify this situation, International Radiotelegraph Convention was held in Washington on *February 18, 1927*, discussed and proposed the Radio act of *1927* recommending the establishment of Federal Radio Commission that supersedes the Radio Act of *1912*. After many failed attempts, the Radio act of *1927* was passed by the congress and US President Coolidge signs the Radio Act of *1927* on *February 23, 1927*, that empowered Federal Radio Commission (FRC) to deal with licensing. The sudden growth of radio station reflects the value of resource like electromagnetic spectrum.

Although there was no specific mention about the regulation of broadcasting network chains, commission can exercise its control over the individual stations. FRC took some immediate actions during *1928* and proposed the closure of 164 stations during *1928*, however after hearing 83 stations were deleted. FRC also made observations regarding relaying too much music on radio stations and cited WCRW Chicago in this regard. FRC also issued a permission to WLW for 500kW operation using call sign W8XO. From *December 19, 1933* FRC also promoted high fidelity stations to operate on 1530, 1550, and 1570 kHz waves and on *May 25, 1934* FRC started issuing call signs for high fidelity stations like W1XBS, American-Republican, Waterbury, Connecticut; W9XBY, First National Television, Inc., Kansas City; W6XAI, Pioneer Mercantile Co., Bakersfield, California; W2XR, John V. L. Hogan, Long Island City, New York.

The FRC was also discredited for inconsistent policy lacking strong regulatory actions by many experts. The broadcasting alliance led by RCA Westinghouse and

General Electric was able to convince FRC to continue with the earlier regulatory framework that was drawn under Commerce Secretary. This led to a lot of criticism of FRC about equality service standards that further led to an amendment called "Davis Amendment" about equality service standards. However, this act became a basis for future Communications Act of *1934* to establish a permanent commission called "The Federal Communications Commission (FCC)" focused on the effective use of electromagnetic spectrum. Later on, Radio act of *1927* was replaced by the US Federal Communications Act of *1934* that was signed by President Franklin D. Roosevelt on *June 19, 1934*. Federal Communication Commission was responsible for breaking up of monopoly of National Broadcasting Company.

6.7.17   Rise of Apex Broadcasting Stations

Early radio receivers operating on the principle of amplitude modulation were hounded by "Static Problems" or "Static Interference" leading to external noise caused by source like thunderstorms, electrical equipments, operation of electric motors, etc., and to yield better results, higher frequency operation was desired. Radio stations were restricted within a band of frequencies from 550 to 1500 kHz and depending on frequency range stations were sorted to various classes called as Class-I, Class-II and Class-III stations. The class-III stations operated in the highest frequency range of 1420 kHzs–1460 kHz.

In *1927*, Federal Radio Commission (FRC) decided not to allow the use of frequency band from 1500 kHzs–2000 kHz, however, another decision was also made to allow the use of some discrete frequencies in a band 1500 to 1600 kHz, and allowed the frequencies 1530, 1550 and 1570 kHz for creating high fidelity amplitude modulated AM stations called "Apex Stations". These stations were provided with 20 kHz separation between different stations while 10 kHz effective separation was provided amongst the neighbouring stations, one operating at higher and other at lower frequency as compared to usual 5 kHz severation.

6.7.18   Lieutenant James H. Doolittle's First Flight Using Aircraft Instrument Guidance in September 1929

On September 24, 1929, Army Lt. James H. Doolittle was the first pilot to take off, navigate and land the aircraft using aircraft navigational aids (Fig. 6.55b–d) [184] in which radio marker beacons were used to indicate his distance from the runway and altimeter was used to know the altitude of the aircraft, while aircraft's orientation in the form of altitude was controlled using directional gyroscope with artificial horizon. The fight started and ended at the Army's Mitchel Field, Garden City, on Long Island (lies in the southeastern direction of New York) took a total time of fifteen minutes and

(a)   (b)

(c)   (d)

Fig. 6.55 First blind flight by Lieutenant James H. Doolittle **a** NY-2 Husky aircraft Courtesy: Air Force Historical Foundation(AFHF)/National archives **b** The Sperry Horizon Courtesy: Aviation Ancestry **c** Kollsman Altimeter Courtesy: Aero Antique **d** Sperry Directional Gyro, Courtesy: Aviation Ancestry

was cooperative efforts of Guggenheim Fund's Full Flight Laboratory, U.S. Army Air Corps, U.S. Dept. of Commerce, Sperry Gyroscope Company, Kollsman Instrument Company and Radio Frequency Laboratories. The radio instruments were installed by Sperry Gyroscope Company afterwards became the part of Lockheed Martin, Kollsman Instrument Company and the Radio Frequency Laboratories (which is now called "Aircraft Radio & Control, a division of Cessna"). The aircraft used was a specially instrumented Army Air Corp NY-2 Husky aircraft (Fig. 6.55a).

6.7.19   Formation of RCA-Victor Company in October 1929

With the progress of wireless telephony and broadcasting business, people also put forward their efforts in the direction of recording of the sound so that it can be played at later instances. The first instance of sound recording was reported as early as in *1857* by a French inventor Edouard-Leon Scott de Martinville using a device called "Phonautograph". Thomas Edison made his first experiment of sound recording at his Menlo Park laboratory using wax coated paper and his device was a modified version of "Phonautograph" called as "Phonograph", the sketch of this device to be used for fabrication of this device was made by Edison on *November 29th, 1877* and by using this sketch, Edison's machinist made complete prototype on *December of 1877*. Using this device, Edison was able to record the recognisable words "Mary had a little lamb, its fleece was white as snow, and everywhere that Mary went, the lamb was sure to go."

Emile Berliner pushed this technology forward to the new heights further, Emile Berliner applied the patents of a cylindrical device called "Gramophone" on *May 4, 1887* and flat-disk record and player patent on *26 September 1887*, however, the mechanism for playing the recorded disks was a problem. Eldridge Johnson, who was working as an operator in a small machine shop in Camden, designed a mechanism that gave a constant speed of 70 rotations per minute which proved to be a true musical player. Berliner merged his company with that of Eldridge Johnson's to form "Victor Talking Machine Company" that mass produced gramophones and records with a well known trademark of dog listening to "His Master's Voice" (HMV). After peak performances, fortunes of "Victor Talking Machine Company" declined and in *October 1929*, Victor Talking Machine Company was purchased by RCA for $154 million and RCA-Victor Company was formed with David Sarnoff as its President which started manufacturing radios and phonographs at New Jersey, USA.

RCA-Victor Company was engaged in manufacturing of RAA series of radio receivers for U.S. Navy in *1931*. The initial versions RAA to RAA-2 were produced till *1934* and RAA-3 sets were manufactured by a division of RCA called RCA Manufacturing Company, Inc. in *1935*. RAA tuning range was from 10 to 1000 kHz divided in five bands as Band-1:10-25 kHz, Band-2:25-63 kHz, Band-3:63-158 kHz, Band-4:158-400 kHz and Band-5:400-1000 kHz.

6.7.20 Commercial Shipboard Wireless Communication Radio Receivers

In 1918, Louis A. Hazeltine designed SE-1420 [185–187], a Classic Shipboard Wireless Receiver and initial contracts for producing SE-1420 were given to Amateur Radio Research and Development Corporation (AMRAD), Wireless Speciality Apparatus (WSA) and Sperry Gyroscope in 1919. In 1920, the cross licensing GE-RCA group, allowed WSA to offer SE-1420 in the form of commercial IP-501.

From the period *1921* to *1923*, Wireless Speciality Apparatus (WSA) which was also a member of cross licensing "Radio Group" built some radios for RCA. Wireless Speciality Apparatus (WSA) became Radiomarine Corporation of America (RMCA), produced shipboard radio receivers for Radio Corporation of America (RCA). IP-501-A LF and MW Shipboard receiver was based on these shipboard radio receivers consisted of three-circuit tuner with a regenerative detector and two stages of transformer coupled audio frequency amplifiers, had exceedingly best performing Antenna Tuner section was entirely isolated and shielded from the main receiver

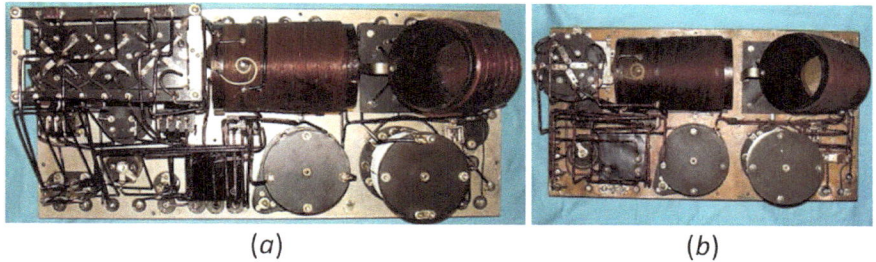

(a)   (b)

Fig. 6.56 Insides of **a** RMCA IP-501-A of 1923 and **b** WSA SE SE-1420 Receivers, Courtesy Radio Boulevard: Western Historic Radio Museum maintained by Henry Rogers

and it was very easy to tune with career wave (CW), when it was operated as autodyne detector. Its secondary tunning was calibrated in meters and had six frequency ranges from 1000 to 40 kHz. This commercial Shipboard Receiver of _1923_ is shown in Fig. 6.56.

Cyril Frank Elwell founded Federal Telegraph Company at Palo Alto, California in _1909_ for the developments of radio communication where Lee De Forest worked from _1911–1913_ on first vacuum tube amplifier and oscillator after the invention of Audion. This company was finally even though it was merged with Clarence Mackay Companies in _August 1927_, still retained its identity. When International Telephone and Telegraph (ITT) purchased Mackay Companies in _1928_, Federal Telegraph Company was still able to retain its existing manufacturing identity. Federal Telegraph Company, Newark, N.J, USA, launched Type 105-A commercial shipboard receiver with serial number _32_081 with first two digits that represented the year _1932_. It consisted of four cathode-type tubes with frequency coverage from 1500 Khz down to 16 kHz in seven tuning ranges (VLF, LF and MW frequencies) The photographs of top and bottom views of this receiver called Pre-World War II Shipboard receiver is shown in Fig. 6.57. During pre-world war II period, the Radiomarine Corporation of America (RMCA) a division of Radio Corporation of America (RCA), was a major supplier to RCA, that was responsible for building communication equipments for communication stations and shipboard installation. In 1938, RMCA introduced, a radio receiver AR-8503 for ship installations [185] and preferably operated on the RM6 battery pack supplying 6V for valve filaments, +22 V for detector and +90 V for plate voltage for the amplifiers. In fact, the US Navy issued a contract for few AR-8503 receivers.

6.7.21 Exploitation of Superheterodyne Receivers

Many entered in the business of manufacturing of superheterodyne radio receivers for general public entertainment, shipboard applications, radio receivers for airport and airway Communication applications and for military applications.

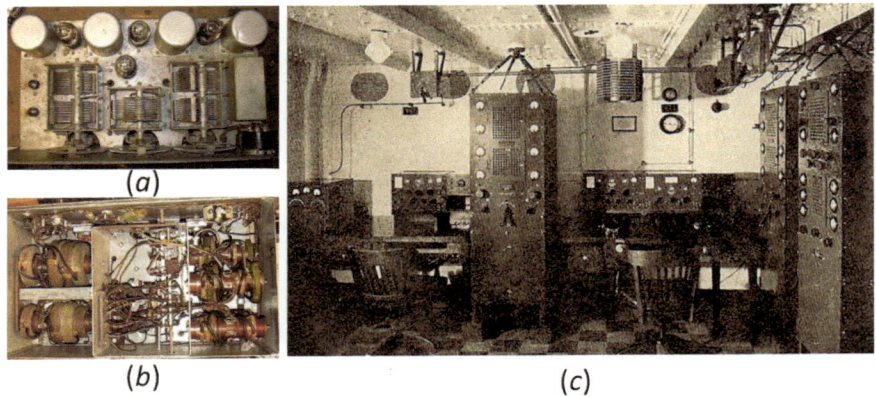

Fig. 6.57 Details of Shipboard receiver Model 105-A built by Federal Telegraph Company for Mackay Radio & Telegraph Company in 1932 operated and tuned in VLF, LF and MW ranges of frequencies from 15 kHz up to 600 kHz. **a** Top view of board inside shipboard receiver, **b** Bottom view of the inside board of shipboard receiver **c** The radio room onboard the S.S. Manhattan dating back from around 1935, two 105-A receivers are shown in the photo along with a 104-A shortwave receiver and Mackay transmitters, *Source* George Sterling's Radio Manual (Third Ed.), Courtesy: Radio Boulevard: Western Historic Radio Museum maintained by Henry Rogers

6.7.21.1   Use of Superheterodyne Receivers for Airports and Airways

Department of Commerce of USA was also looking after airports and airways, contracted three radio receiver companies for upgradation of airports and airways on the basis on three different aspects called ground-based airport radio receivers, new airborne gear and upgradation of existing transmitters for entire systems for airports and airways. In late 1931, the Department of Commerce selected National Company Inc. to replace old regenerative sets with a new type of superheterodyne receivers. The contract identified by number 32–15305 was awarded to National Company on May 12, 1932, for ground-based airport radio receivers, Aircraft Radio Corporation (ARC originated in 1924 in Boonton, New Jersey was set up as a wholly owned subsidiary as a radio division of Radio Frequency Laboratories (RFL)) was awarded a contract for the new airborne gear while the contract for supplying transmitters for entire system upgradation of airport and airway was awarded to General Electric Co.

Ground-based airport radio receivers:
In 1927, National Company Inc. started the development of shortwave receivers under the direction of James Millen called "Thrill Box".

Due to the award of contract from the Department of Commerce, USA, on May 12, 1932 National Company, Inc., first time it entered into superheterodyne market and started building ground-based RHM type shortwave superheterodyne receivers (Fig. 6.58)(a) for airports using nine tubes that became its trademark design for ground-based applications in the airports also known as "RHM commercial airport

(a)   (b)

Fig. 6.58 Ground and airborne radio receivers **a** The original ground station based RHM type radio receiver originally manufactured by National Co. Courtesy: Radio Boulevard: Western Historic Radio Museum maintained by Henry Rogers **b** Airborne Radio Receiver RA-1B used in the Navy aircraft Courtesy: AAFRadio, aafradio.org a sight maintained by Michael Hanz

receiver". The other receivers manufactured by National Company, Inc. were AGS-X Single Signal Receiver in March 1933, RHQ receiver, RIO series receiver (also identified as AGL) in 1933 for navigational and communications requirements, HRO receiver in October 1934, RCR Moving Coil Airport Receiver in 1937.

Airborne radio receivers:
Aircraft receiving equipment type RA1-B also called as airborne radio receiver is shown in Fig. 6.58b, manufactured by Bendix was compact and very popular with commercial airlines and was manufactured even after the war. The set probably came into use in commercial service immediately prior to WW2.

6.7.22   Radio Wars

In *1926*, the total energy transmitted by radio waves amounted to 116 kW which was raised to 8000 kW in *1938* leading to the race of increasing transmission power leading to a radio war of broadcasts and counter broadcasts with intension for jamming other transmitters working at same frequencies. The good example of rivalary that can be cited as the race between Prague and Hungary resulting in the domination of language wars. The jamming methods adopted were Morse code, Disturbing voice on near same wavelength, etc. Nazi's made people to listen to Nazi broadcasts restricting them to watch foreign programmes. Shortwave stations in Europe increased many folds from three in *1930* to forty on *1937*. Taking the advantage of Berlin Olympic games in *1936*, Germany added eight short wave stations for its own propaganda to reach Africa, south and north America. It became a trend to arrange broadcast in many languages to influence other language speaking countries. Form *1937* BBC and NBC started broadcasting in other languages using their short wave radio service.

6.8   Conclusion

The remarkable contribution of Reginald A. Fessenden in the field of wireless telegraphy generated a great deal of interest amongst amateurs sparked by the rapid developments of wireless radio receivers, and Hugo Gernsback was the first to offer affordable wireless radio equipment kits for experimenters. American Radio Relay League (ARRL) founded by Hiram Percy also came to the rescue of amateurs. Although initial systems were successfully working, suffered from many drawbacks and needed amplification. An invention of thermionic devices capable of signal detection and amplification made a timely entry in the telephony systems that made a drastic difference to the existing systems.

Chapter 7
The Entry of Television

Abstract This chapter is dedicated to the developments in the area of television before World War II. It all started with still image sensing and duplicating it on another device connected by the wires, involved picture scanning, generation of picture signal for transmission and reception of picture signal for reproduction of originally transmitted picture. The Pantelegraph, a precursor of modern facsimile machine, senses still image and transmits it to an another location. Alexander Bain patented the principles of facsimile machine; however, he could not take an advantage of his invention as Frederick Collier Bakewell took a patent of his superior device earlier than Alexander Bain. The world's first practical operating facsimile machine called "Pantelegraph" was developed by Abbe Giovanna Caselli and was first person to transmit a still image over wires. The landmark discovery of photosensitive selenium by Swedish chemist Jöns Jacob Berzelius paved a way for non-contact sensing of images. Arthur Korn was the first person to use photosensitive properties of selenium in his fax machine. Édouard Belin improved the methods of Arthur Korn and added wireless transmission of pictures using radio waves and photoelectric methods and was able to measure intensity of each point of picture known as pixel. Constantin M. SENLECQ was first to use selenium in his first device called Télectroscope and George R. Carey used array of array of selenium and used current variation depending on intensity of light in sensing the image while English physicist Shelford Bidwell was first to use a photocells for scanning of pictures. William Lucas contributed to a scanning mechanism used in modern television. Paul Nipkow introduced a spiral scanning mechanism used in early televisions. Many inventors like John P. Gassiot, Julius Plücker, Cromwell Fleetwood Varley and William Crookes contributed to the developments of cathode-ray tubes used in reproduction of moving images. Eugen Goldstein introduced a concept of perforated grid to create canal rays called "cathode rays". Karl Ferdinand Braun introduced cathode-ray tube with fluorescent screen also named after him called "Braun tube". J.J. Thomson discovered that beam of cathode rays consists of lightest negatively charged particles called corpuscles called "Electrons".

It was a time now to introduce a term "television" that was introduced by Constantin Persky while Max Dieckmann and Gustav Glage introduced a concept of raster scanning in television to reproduce images on CRTs. St. Petersburg, Boris Rosing developed a system using mechanical mirror drum scanner for transmission of crude images to receive them in cathode-ray tube called "Electric Telescope". Alan Archibald Campbell-Swinton describes how distance vision can be achieved. It was the first time that John Logie Baird transmitted the pictures of hands of Victor Mills using wireless transmission called "Seeing by wireless". Baird also held the public demonstration of Silhouette images on his device "Silhouette Television" while On October 02, 1925, John Logie Baird transmitted moving image with variation in shades while on January 26, 1926 Baird practically demonstrates his television first time to the members of Royal Society and further on 1928 transatlantic television signal from London to Hartsdale, New York was sent by Baird Television Development Company.

In the beginning of 1924, Kálmán Tihanyi introduced charge-storage technology the was used for scanning in camera tube that facilitated the developments of all electronic television system called "Radioscope". The demonstration of true all electronic television by sending moving pictures was given by Philo Taylor Farnsworth in San Francisco on Sept. 7, 1927. On July 03, 1928, John Logie Baird demonstrated world's first colour transmission. After the developments of all electronic television, there was a race for improvement in resolution of television systems. The experimental broadcasts were started by WGY of Schenectady, 9XAA of WCFL also known as W9XAA Chicago, Charles Francis Jenkins worked regular schedules of broadcasting using Call sign 3XK and on WRNY, New York and 2XAL along with WRNY went on air on August 13, 1928 and then RCA and BBC started their TV transmission.

7.1 Work in the Area of Television Broadcasting

Television broadcasting is thought to be simultaneous transmission of both series of picture frames with incremental changes in them along with the sound to produce the dynamic effects of movements of objects in the picture frame called motion pictures, hence people started thinking of transmission of picture frames with incremental changes and sound from one place and simultaneous reception of both sight and sound at some other places. The developments in sound transmission and reception were in much more advanced stage and had an edge over to picture transmission and reception.

7.1.1 Transmission and Reproduction of Still Images

It was a time to evolve the methods of picture transmission from one place and receive it at some other place. The picture transmission and reception had three components called scanning, generating picture signal for transmission, reception of picture signal for reproduction of originally transmitted picture.

Initially people started experimenting by sending still image from one place to other and resultant machines or devices were called "Pantelegraph" which was a precursor of modern day facsimile machine named by combining pantograph and telegraph and later it was called fax machine.

7.1.1.1 Direct Contact Sensing of Still Images

It all started with duplicator device that transmits the copy by wire or radio called facsimile machine. A Scottish instrument maker Alexander Bain patented the principles of facsimile machine [188] between *1843* and *1846*. Bain used mechanical movements clock pendulum to synchronise line by line scanning of messages and used metallic pins mounted on insulating drum that acted as electric probes producing pulses for sensing the picture. The sensed pulses were transmitted by wire to reproduce the original picture at receiving end using electrochemically sensitive paper. He applied a patent for his device in 1850 but it was too late to take the advantage of situation as Frederick Collier Bakewell took a patent of his superior device earlier than Alexander Bain.

Alexander Bain's facsimile machine concept was immediately extended by an English physicist Frederick Collier Bakewell and designed a copying electric telegraph (Fig. 7.1) and first successful telefax transmission occurred in *September 1847* [189] by sending image telegraph messages between Seymour street in London and Slough. His copying electric telegraph called "Chemical Telegraph" was patented vide British Patent No. 12352 dated *June 02, 1847* while public demonstration was held at an exhibition in London in 1951.

The world's first practical operating facsimile machine with synchronisation between sending and receiving apparatus called Pantelegraph was developed by Abbe Giovanna Caselli and its demonstration was held on *May 10, 1860*. Giovanna Caselli was the first person to transmit a still image over wires.

Many researchers like Ernest Hummel (*1895*), Arthur Korn (*1902*), Edouard Belin (*1914*), Alexander Muirhead (*1947*), etc., worked and contributed to telefax developments using various principles like electro chemical, electro magnetic, Light rays, photoelectric cells, etc.,

7.1.1.2 Photoelectric Sensing of Still Images

It was a time for contactless sensing of picture. In *1817*, a Swedish chemist Jöns Jacob Berzelius made landmark discovery of light-sensitive properties of selenium and its utility to picture sensing remained unexplored till *1873* when a telegraph operator Joseph May in Valentia, Ireland, discovered accidently that selenium rod exposed to sunlight change its resistance that was further confirmed by Willoughby Smith. Both Joseph May and Willoughby Smith experimented with selenium to

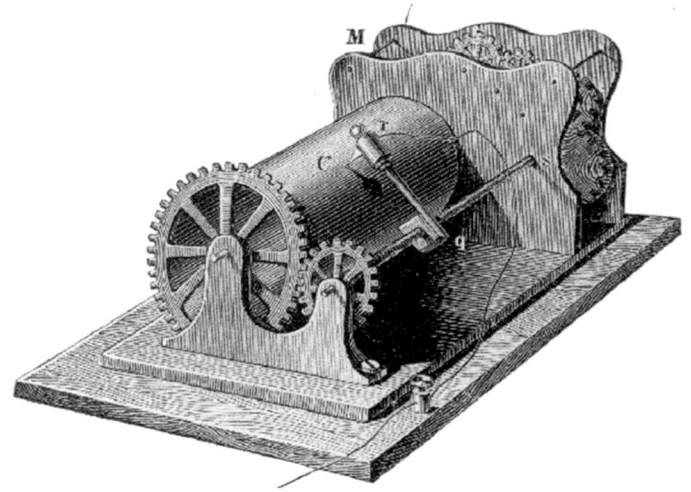

Fig. 7.1 Bakewell's Fascilmile Device, *Credit* CC BY-SA 3.0, European Patent Office |January 1, 1848|, Bakewell improved facsimile 1848

figure out that it could be used for transforming images in to electrical signal which could be transmitted using wires. In *October 17, 1906*, German mathematician and physicist, Arthur Korn unlike his predecessors used photosensitive selenium in his fax machine and transmitted a photograph of Crown Prince William over a distance of approximately 1120 miles.

7.1.1.3   Photoelectric Sensing and Wireless Transmission of Still Images

Arthur Korn's methods were further improved by Édouard Belin in *1921* by adding wireless transmission of pictures using radio waves and photoelectric methods by measuring intensity at each point of picture using "electric eye", the device that detects the presence and absence of light.

7.1.2 *Transmission and Reproduction of Moving Pictures*

The facsimile devices used for still photographs pioneered on the basis of mechanical scanning of the images were restricted in a particular frame. After the inventions of still image transmission and reception techniques, it was an era of transmission and reception of moving images or pictures based on series of overlapping frames with incremental changes to cerate effect of moving picture.

7.1.2.1   Photoelectric Sensing

In *1879*, Constantin M. SENLECQ was first to use selenium in his first device called Télectroscope [190, 191] in which changing electrical conductivity properties of selenium when exposed to light were used. In *June 5, 1880*, George R. Carey used the array of selenium photocells [192] and current through each photocell varies depending on intensity of light in that particular part of image to sense the image which is reproduced by the receiver at other end, in which chemically sensitive paper placed between respective contacts produces an image on paper. The mechanism was static and required array of photocells and with individual connections to photocells making it a jugglery of wires. George R. Carey further modified his mechanism that used single selenium photocell to scan the image by moving it over entire image which required mechanical scanning involving moving parts.

An English physicist Shelford Bidwell was first to use a photocells for scanning of pictures as reported in an article entitled "Tele-Photography" published in *February 10, 1881* issue of "Nature".

William Lucas, in *1882*, made significant contributions in horizontal and vertical scanning that is used in modern television.

In *1884*, Paul Nipkow introduced the concept of spirally perforated discs for scanning of the images called Nipkow disk, for which he applied a patent to Berlin patent office on "electric telescope for the electric reproduction of illuminating objects" that was granted in *January 15, 1885*, however, it was not known whether he actually constructed it.

7.1.2.2   Developments in the Area of Cathode-Ray Tubes

Cathode-ray tubes form the basis of reproduction pictures from the received video signal, work on the basis of electric discharge. In *March 04,1858* John P. Gassiot [193], vice president of the Royal Society in his Bakerian lecture, reports deflection of electric discharge by means of magnetic and electrostatic fields.

The German mathematician and physicist Julius Plücker [194] while working as professor at University of Bonn in *1859*, studied electric discharge in rarified gases using vacuum tubes of his Bonn colleague Geissler under the influence of magnetic field, discovered the formation of fluorescent glows on the glass walls of vacuum tubes and later on it was identified that glows were produced by cathode rays.

This work was continued by another German physicist Johann Wilhelm Hittorf [195] in *1869*, and pupil of Julius Plücker while working at University of Münster, he worked on the various aspects of passage of electricity through various gases. He made the tubes with absolute vacuum and ascertained that the rays originating from negative electrode when hit the glass wall produce fluorescence.

In *1871*, Cromwell Fleetwood Varley's paper [196] suggested that cathode rays consist of streams of electrical particles caused by collision of particles, because these particles were deflected in presence of magnetic field.

William Crookes experimented with the current flow between two electrodes mounted in vacuum tubes [197, 198] when high voltage is applied between two electrodes. Since filament is absent, negative electrode terminal called cold cathode and positive electrode called anode are applied with high voltage and electrons are generated due to ionisation of air due to applied high voltage and the positive ions are attracted by cathode and when positive ions strike the cathode material, leading to knocking out of large number of electrons which are repelled by cathode and attracted towards anode creating a beam of lightest particles called electrons towards anode called cathode rays. William Crookes confirms this existence of beam of lightest particles (cathode rays) on _1879_.

Eugen Goldstein [199, 200] worked on perforated cathodes and work based on perforated cathodes appeared in the form of two papers in _1876_ and later in _1886_ in Monthly Report of the Royal Prussian Academy of Science at Berlin. He observed that canal rays came out of these perforated cathodes. He was responsible for these rays to be called "Cathode Rays".

Heinrich Hertz, after appointment as professor of physics at the University of Bonn in _1889_, pursued his research in the area discharge of electricity in rarefied gases [201, 202] and reported that cathode ray discharge is continuous and not discrete and tried to deflect cathode rays with an electric field, but was not able to do so and hence concluded that cathode rays are not made up of charged particles. He believed that since cathode rays are waves, they could be deflected with magnetic fields.

Philipp Lenard came to Bonn on _1 April 1891_ to work as assistant to Hertz. Taking the work of Hertz as a base, in _1892_ he constructed tube with "Lenard window", able to direct the rays out of the discharge space. Although all of these persons called Jean Perrin, Willy Wien and J. J. Thomson had similar opinion about cathode rays consisting of negatively charged particles, Philipp Lenard in his _1898_ publication [203] showed the dispute with J.J. Thomson and claims made by him.

Wilhelm Roentgen [204] in _1895_ experimented with cathode rays and their effects on coated surfaces producing different types of rays. He experimented with a paper plate coated with barium platinocyanide that became fluorescent. He also received the first Nobel prize in _1901_ in Physics.

In _1896_, Karl Ferdinand Braun [205] introduced cathode ray tube with fluorescent screen which was highly useful for measuring waveform properties and evolved instrument was called cathode-ray tube oscilloscope. For his work, Karl Ferdinand Braun shared Nobel Prize with Guglielmo Marconi in _1909_.

On _April 30, 1897_, Joseph John Thomson discovered that beam of cathode rays consists of lightest negatively charged particles called corpuscles [202] but later scientists preferred it's name as "electron" that was suggested by George Johnstone Stoney in _1894_, prior to Thomson's discovery. Thomson also experimented with electrical deflection of cathode rays by means of electrostatic field applied to parallel plates making it useful for creating images on the glass wall opposite to cathode location leading to principle creation of images using electrical signals.

The term _"**Television**"_ was coined first time by Constantin Perskyi in his paper that carried the review of electromechanical technologies and work of Nipkow, pre-

Fig. 7.2 The concept of
raster scanning of image
used in cathode-ray tube

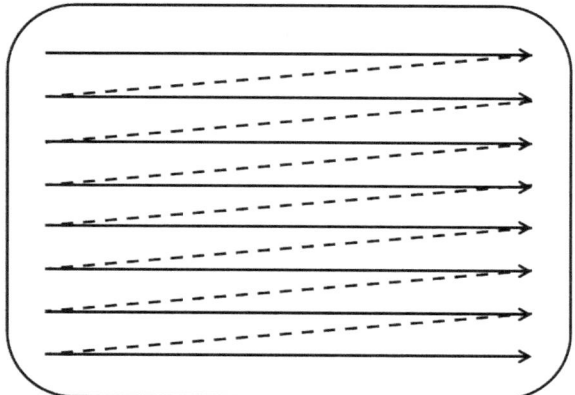

sented at International Electricity Congress at the International World Fair in Paris
on *25 August 1900*.

7.1.2.3 Raster Scanning

Max Dieckmann and Gustav Glage were first to use raster scan principle to produce image on cathode-ray tube in *1906* wherein image was divided in to successive
samples called "pixels"/indexPixels along scan lines and each scan line is read and
transmitted simultaneously. In modern-day television systems, each scan line consisting of a row of pixels can be stored in memories. The principle of raster scanning
is shown in Fig. 7.2.

7.1.2.4 Use of Cathode-Ray Tubes in Television

In St. Petersburg, Boris Rosing, a Russian scientist along with his student Vladimir
Zworykin developed a system using mechanical mirror drum scanner for transmission of crude images to receive them in cathode-ray tube (CRT) or Braun tube which
acted as a receiver where the transmitted images can be viewed. Noting the short-
comings of mechanical scanner, he thought that electrical signal related to images
can be directly produced using static mechanism instead of mechanical scanner using
CRT and photocell combination, and electrical signal produced can be used for transmission purpose. For this type of system, he filed a patent application in Russia on
July 25, 1907 called "Electric Telescope".

In *June 04, 1908*, a Scottish electrical consultant Alan Archibald Campbell-
Swinton (FRS), and member of Roentgen Society [206], describes about how "Distant Electric Vision" [207] can be achieved using CRT in a paper published in scientific journal "Nature", provided theoretical basis for completely electronic television

Fig. 7.3 Baird's Television
using Nipkow scanning disk,
Courtesy: CC BY-SA 3.0, H.
Winfield Secor, |August 13,
2014| "John Logie Baird
speaks his mind" in
Television News magazine,
Popular Book Corp., New
York, Vol. 1, No. 6,
January–February 1932, p.
410 on
AmericanRadioHistory.com
website

system. He was able to synchronously deflect beams of cathode-ray tubes of transmitting and receiving stations by varying fields of two electromagnets placed at right angles energised by two AC currents of wide range of frequencies. This work laid down the foundations of practical television.

During _1918–1919_, John Logie Baird decided to work on a device based on Paul Nipkow's Disk scanning disk and first time in _February 1923_, John Logie Baird, transmitted the pictures of hands of Victor Mills. On _June 27, 1923_, John Logie Baird put an advertisement Times's personal column called "Seeing by Wireless" inviting the assistance from others and also patented his device "System of transmitting views, portraits and scenes by telegraphy or wireless telegraphy" Patent No. 222604 dated _July 26, 1923_. In _April 03, 1924_, F.H. Robinson, reporter, wrote about Baird's work on mechanical radio vision device highlighting its readiness for commercialization. Baird held the public demonstration of Silhouette images (dark shadow images with outer lining matching with objects or subjects in motion and interior of such images are featureless) in motion on his device "Silhouette Television" at Selfridges store on Oxford Street in London _March 25, 1925_.

On _October 02, 1925_, John Logie Baird transmitted moving image by variation in shades. On _January 26, 1926_, Baird practically demonstrated his television first time to the members of Royal Society at his premises at his laboratory at 22 Frith Street in the Soho district of London where at least 50 scientists visited in batches and this was said to be the world's first public live television broadcast demonstration [208] based on rotating mechanical scanning disk is shown in Fig. 7.3. His apparatus was called "Televisor" for which he was awarded with two patents No. 289,104 and 292,632 on _October 1926_ and _January 1927_, respectively.

In _1927_, Baird transmitted television signal over telephone lines between London and Glasgow at the distance of 438 miles while on _1928_ first transatlantic television signal from London to Hartsdale, New York by Baird Television Development Company.

Charles Francis Jenkins a farm boy born near Dayton, Ohio, took a keen interest in radio technologies and developed an apparatus called "Teloramaphone" which sent pictures to distant point was a source of news during those days. Interest of his apparatus grown amongst ham operators, and he was then well associated with ham operators organisation called ARRL which had a great role in popularising radio broadcasting. Jenkins anticipated that ARRL can also play a great role in television broadcasting, and ARRL members helped him to compete with growing number of broadcasting companies. G.L Bidwell, member and Director of Atlantic division, in *1925* quoted that "motion picture by radio are here!"

In the beginning of *1924*, Kálmán Tihanyi introduced charge-storage technology for scanning in camera tube, and using this technology, he introduced fully electronic television system for both scanning and displaying for which he applied for patent in Hungary in 1926 called "Radioscope".

During *1925*, Japanese Kenjiro Takayanagi started his research in television using Nipkow disk scanning to generate electric signal, however, unlike Baird, he used cathode-ray tube (CRT) to display received signal making it a all electronic television and in *December 25, 1926*, Kenjiro Takayanagi was successful in demonstrating his all electronic television system at Hamamatsu Industrial High School in Japan.

Soviet engineer Boris Pavlovich Grabovsky claimed to have made first electronic TV broadcast in *1926*, however, many do not substantiate or subscribe to his claim.

AT&T held first highly publicised public event of demonstration of television that was held by Herbert Ives and his research team on *April 17, 1927* in which images were transmitted over a distance of 200 miles between Washington DC and New York with brief appearances of key personalities like Herbert Hoover, Secretary of Commerce and John J. Carty, AT&T vice-president.

The demonstration of sending moving pictures from one place to other place was first time demonstrated by Philo Taylor Farnsworth in San Francisco on *Sept. 7, 1927* when he captured moving pictures, coded on radio waves and recaptured back and reproduced on the screen at his laboratory at 202 Green Street in San Francisco. It was true electronic television system without any use of Nipkow disk. In *1929*, the system was further improved and generator with moving parts was also removed to make a system completely static.

In *November 1929*, Vladimir Zworykin a student of Rosing experimenting with improved cathode-ray tube for receiving pictures filed a patent on a device called "Kinescope". He also met David Sarnoff of RCA who hired him for development of television at RCA laboratories in Camden, New Jersey.

7.1.2.5   Birth of Colour Television

On *July 03, 1928*, John Logie Baird demonstrated world's first colour transmission using mechanical scanning disks at both transmitting and receiving ends using three spirals of holes on them. Each spiral has filter of three primary colours operated on the principle of stereoscopic vision.

Fig. 7.4 *Source* CC BY-SA 3.0, Boston Public Library |February 2011| Flickr: WGY broadcasting station, Schenectady, N. Y.

7.1.2.6 Initial Experimental Television Broadcasting Efforts

Initial television broadcasting was in the form of experimental test broadcasting using existing call signs and in *May 10, 1928*, station with call sign WGY operating on 790 kHz (AM), (where W stands for wireless, G stands for General Electric and Y stands for last letter of Schenectady where it is located), licensed to Schenectady, New York, started its television tests using 24 line resolution mechanical system designed by Ernst Alexanderson of GE. The televised drama program was based on "Queens Messenger" written by J. Harley Manners. WGY station is shown in Fig. 7.4.

The first TV station in Chicago went in operation on *June 19, 1928* was 9XAA operating on the frequency 4800 kHz (short-wave station of WCFL, IL. Chicago officially W9XAA).

By *July 02, 1928* Charles Francis Jenkins [209] worked a regular schedule for broadcasting radio movies for hams and various interested movie watchers with short-wave receivers and broadcasted movies from his research lab using his experimental station with call sign 3XK operating on 46.72 m (6416.80 kHz). This transmission was called silhouettes or shadow pictures or shadow graphs.

Further two stations called WRNY (920 kHz), New York, and 2XAL (9700 kHz) along with WRNY went on air on *August 13, 1928* for mechanically scanned TV transmission.

7.1.2.7 Race for Improving Resolution

In *1931*, John Logie Baird decided to work ultra-shortwave transmission as medium waves used by BBC, limited Baird's transmission to only 30 lines. He gave demonstration of his ultra-shortwave transmitter at Long Acre on *April 29, 1932*. In *1932*, due to shortage of funds, Baird Television Ltd (BTL) was taken over by Isidore Ostrer and the Gaumont-British Film Corporation. One year later (*1933*), boardroom coup relived Baird from his duties on pretext of being contracting on mechanical TV rather than fully electronics TV; however, due to his fatherly contribution in TV, he was retained as managing director and Captain A.D.G. West who was engineer with BBC and Isaac Shoenberg's EMI active in all electronic TV became technical director. Captain A.D.G. West worked on increasing the resolution of mechanical rotating disc scanning system in vacuum. In *July 1933*, BTL was then moved to new location at The Crystal Palace in Sydenham, South London, where ultramodern facilities were set up for manufacture of TV station and receivers and Baird was completely expelled from BTL and BTL was then free to develop new technologies of all electronic TV technologies. It was a competition between BTV and Shoenberg's EMI; however, Shoenberg's EMI took a lead in demonstration of 120 line all electronic TV system superior to BTL to BBC and GPO in early *1934* and by *April 18, 1934*, Shoenberg's EMI team drastically improved the performance of the system using vision mixer with six channels.

7.2 BBC's Entry in TV Broadcasting

Baird's TV broadcast through BBC started from *30 September 1929* and continued till *1932*. After demonstration of 120 line all electronic TV by Shoenberg's EMI team to postmaster general, in Britain, a British government set up a committee headed by Lord Selsdon to identify the relative merits of various television systems developed as well as medium to be used. The committee recommended BBC to start high definition TV and both Baird and Marconi-EMI were invited to supply equipment for the experimental site for which minimum acceptable standard was set as 240 lines with 25 picture frames per second and both companies failed to satisfy the recommended standard reliably. However, Baird-Fernseh AG system of film scanning worked well at lesser standard of 180 lines.

7.3 RCA's Television Broadcasting Experiments and FCC's Standardisation Efforts

In *1923*, Vladimir Zworykin while working in Westinghouse filed a patent for all electronic television system that used cathode-ray tube for both transmission and reception of images. In *1924*, he started building television system based on his patent and demonstration of all electronic television system was held before the executives of Westinghouse in *1925*. However, executives of Westinghouse were not so much interested in his work. At Westinghouse he reassigned a work on "Photoelectric Cells", while in late *1928*, was sent to Europe for gather the information about the joint work being done by Westinghouse and RCA on television research, where he got impressed by the work of Fernand Holweck and Pierre Chevallier and the cathode-ray tube designed by them called "Holweck-Chevallier tube" that used electrostatic field to focus electron beam. Based on this visit, he engineered some great ideas in his mind which he did not share with Westinghouse executives. However, for implementation of his ideas, as per the advise of Sam Kintner, vice president, in *January 1929*, Vladimir Zworykin met David Sarnoff who persuaded Westinghouse to provide necessary resources and by the end of *1929* he perfected cathode-ray receiver called "Kinescope", but still used mechanical device called spinning mirror in transition part of his apparatus. Using Kinescope Vladimir Zworykin received experimental television signals at his home from Westinghouse's KDKA, radio station in Pittsburgh. In *1930*, Westinghouse's television research was transferred to RCA's laboratory at Camden, New Jersey, and Vladimir Zworykin became head of television division at RCA's Camden laboratory.

Vladimir Zworykin now focused attention to remove mechanical parts in transmission side and in *April 1930*, he visited Philo Farnsworth's laboratory in San Francisco where he got inspired by the Farnsworth's transmission tube called "Image Dissector", and developed improved camera tube mechanism called "Iconoscope" that was useful for all electronic television system for which he filed a patent in *1931*.

David Sarnoff, President of the Radio Corporation of America (RCA), saw the better future in television business and leased 85th floor of empire state building for experimental television broadcast. RCA through its own subsidiary called National Broadcasting Corporation applied two separate licences for sight and sound to Federal Radio Commission (FRC) on *July 1, 1931* which were issued on issued on *July 24, 1931*. The call sign W2XF for sight channel operating on 44 MHz and 5 kW power (2 kW antenna output) was issued on *December 1931*. While for sound channel, was assigned call sign W2XK for operation on 61 MHz and 2.5 kW power. Both of these channels have associated vertical dipole antennas installed on the top of empire state building.

Nipkow mechanical scanning disk provided video signal to the transmitter, in this method, spirally perforated rotating disk is used to pass the light generated by arc that was sensed by photocells at other end to derive video signal. At receiver, picked up video signals were reproduced by cathode-ray tube "kinescope", which was invented at Camden laboratories of RCA by Vladimir Zworykin. Picture from "kinescope" was optically enlarged and projected on the screen. The moving images were constructed using sequential 24 frames with each frame containing 120 lines.

7.3.1 RCA's All Electronic Television Broadcasting from 1933 to 1936

RCA kept Vladimir Zworykin's iconoscope developments till *1933* and only after *1933* Vladimir Zworykin made the developments of all electronic television public when RCA successfully demonstrated its all electronic television broadcast using its

W2XBS experimental television transmitter station located on Empire State Building. This transmission was using 240 line resolution, 24 frames per second, sequential scanning, Bandwidth of 2 MHz and both Video and audio carriers were full side band amplitude modulated and went up to *1937*.

7.3.1.1 Federal Communication Act of 1934

The federal communication act of *1934* aimed at more effective and comprehensive framework of federal regulations of various types of communication methods covering telegraph, telephone, television and radio communications in the form of various statutes. For effective regulations, this act created Federal Communications Commission (FCC) to regulate all types of communication industries. This act can also be updated time to time as per the provisions of the act, to cover satellite television. FCC was provided powers for regulation of radio spectrum, rates and fees, standards, competition, terms of subscriber access, commercials, broadcasting in the public interest, government use of communications systems. During this period FCC established by an act of congress on *June 22, 1934*, decided to plan the VHF radio spectrum for both experimenters and television usage, the experimenters who were otherwise using any frequency above 30 MHz were shifted to frequencies above 100 MHz and television stations were allotted any frequency in between 42 to 46 MHz and 60 to 86 MHz without any channel assignments.

7.3.2 Invention of Frequency Modulation Principle for Reducing Disturbances in Radio Signalling-1935

Early radio receivers operating on the principle of amplitude modulation were hounded by "Static Problems" or "Static Interference" leading to external noise caused by source like thunder storms, electrical equipments, operation of electric motors, etc., and to yield better results, many stations were working on amplitude modulation of frequencies in VHF range (30 to 300 MHz).

In the year *1922*, J R Carson [210] of Department of Development and Research, American Telephone and Telegraph Company, New York, describes the theory of frequency modulation in his paper entitled "Notes on The Theory of Modulation".

After Major Edwin Howard Armstrong lost a legal battle lasting for 21 years with Lee Forest regarding rights of regenerative circuits in *1922*, Armstrong continued to work on "static problem" plunged in early radios in spite of his colleague's assertion that static problem could not be avoided.

In Amplitude modulation, although physical reach of the signal was better, it resulted in poor signal quality. Armstrong sought to improve the quality of signal by way of changing frequency of carrier signal rather than changing its amplitude and devised a frequency modulated (FM) radio circuit and it was matured enough to be demonstrated and patented. Edwin Howard Armstrong got patent for it on *December 26, 1933* and in the same month demonstration was arranged for RCA staff that included David Sarnoff and further demonstration in continuation was held in *January 1934*. From *May 1934*, Edwin Howard Armstrong starts a work towards experimenting frequency modulated (FM) broadcasting from empire state building and on *June 16, 1934*, Armstrong arranges first test in cooperation with RCA by using their W2XF transmitter installed on empire state building operating on 41 MHz and experiments were conducted for both amplitude and frequency modulations to have a good comparison of both types of broadcasting techniques and to the surprise of all FM broadcast was clean and sound rich and listeners were shocked by the difference of sound quality. On *April 1935*, Edwin Howard Armstrong receives a message from David Sarnoff to stop frequency modulated transmission from empire state building and finally this Armstrong's FM broadcast from empire state building was removed on October *1935*. The main reason for David Sarnoff to remove FM radio transmission was due to his interest in more promising TV business rather than FM radio business.

Based on his invention of frequency modulation principle, Edwin H Armstrong presented a paper entitled "A Method of Reducing Disturbances in Radio Signaling by a System of Frequency Modulation [211, 212]" in New York section of IRE on *November 6, 1935*. He also demonstrated his FM techniques before Institute of Radio Engineers with a physical distance of 17 miles using transmitter with call sign W2AG at Yonkers, New York, and this was his first public demonstration. His paper presented before IRE was published in proceedings of IRE on *May 1936*, while on *June 17, 1936*, FM radio was demonstrated to the FCC for the first time.

7.3.2.1   Starting of Experimental and Non-Commercial FM Radio Stations

On *July 1936*, Armstrong received experimental licence for FM Station. In *1937*, Edwin Howard Armstrong ordered 25 FM receivers from General Electric (GE) at the cost of $400 each for setting up various broadcasting stations.

On *April 10, 1938*, Major Edwin Howard Armstrong started experimental radio station with call sign W2XMN on 43.7 MHz and 600 W power. By *July 18, 1939*, regular programs were aired using 35 kW power. After this success, GE and many other started manufacturing FM receivers that boosted FM broadcasting business.

Edwin Howard Armstrong's former friend and colleague David Sarnoff was not supportive to FM techniques, turned away from his commitment to FM developments. However, John Shepard, a business executive was believed that FM technique could be beneficial and profitable in the business, supported Edwin Howard Armstrong and offered Yankee Network [213] to Armstrong for his experiments to bring FM broadcasting to Boston. In the spring of *1937*, John Shepard applied for FM radio station and was granted a permission by FCC and radio station with call sign W1XOJ, also known by "Static-less Radio" was built on mount Asnebumskit near Paxton, MA,

USA that operated on 43 MHz and 2 kW, become first FM station to go on air on *May 27, 1939*, and later on its power was further increased to 50 kW.

In *1940*, Federal Communication Commission (FCC) decided to allocate spectrum from 2 to 50 MHz for FM broaching consisting of forty FM channels and in *October 1940*, 15 FM stations were given the permissions. The first FM station to begin its operation after decision on FM spectrum was W47NV which became WSM-FM Nashville, Tennessee (USA) that was signed on new years day of *1941*. The first non-commercial station KALW, San Francisco, was licensed on *March 1941*, the transmitter was implemented using a earlier RCA transmission equipment given to SF Unified School District for Golden Gate International Exposition during *1939–40*.

7.3.3 RCA's Television Broadcasting Developments from 1936–1938

Looking at the success of promising television broadcasting, RCA established new television studios linked by radio link at 177 MHz as well as by cable to Empire State Building Transmitter and started conducting extensive filed trails of their television systems from *June 1936* by distributing 100 television receivers to their employees for placing them to their homes and offices. This transmission was using 343 line resolution, 30 frames per second, Interlace(2:1) scanning, Bandwidth of 5.75 MHz and both Video and audio carriers were full side band amplitude modulated.

7.3.3.1 FCC's Informal Hearing of 1936

On *June 15, 1936*, FCC started informal hearing [214] for allocation of spectrum allocation above 30 MHz and the word spectrum standardisation came in forefront. The various associations related to television form a sub-committee for this cause, and the Radio Manufacturers Association (RMA) recommended that the bandwidth of television channel should be at leat 6 MHz which is still valid even today and a resolution of 441 lines was also proposed. The resolution which was seemed to be beyond capacity became reality after eight months after standard was proposed when Philco convincingly demonstrated it on *February 11, 1937*. In this standard, frequencies from 44 to 294 MHz were spilt into 19 channels in which each channel is 6 MHz wide. The channel assignments were 44–50 MHz, 50–56 MHz, ... , and 288–294 MHz. RCA had been quickly assigned a channel No. 1 starting from 44 to 50 MHz. These standards planned to became effective on *October 13, 1938*. The TV channel assignments are shown in Fig. 7.5.

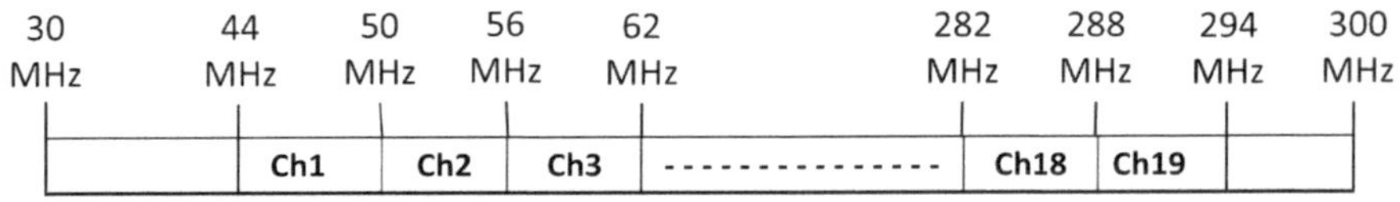

30 MHz	44 MHz	50 MHz	56 MHz	62 MHz		282 MHz	288 MHz	294 MHz	300 MHz
		Ch1	Ch2	Ch3	- - - - - - - - - - - - - - -	Ch18	Ch19		

Fig. 7.5 FCC TV channel standard: the television channels spread over VHF frequency band

7.3.4 RCA's Regular Television Broadcast from 1939

One week after allocation of channels on *October 30, 1938*, RCA announced that regular public television programs will commence from *April 30, 1939* using RCA"s W2XBS radio station in New York, the date that coincided with the opening of Yorks's "World Fair".

7.3.4.1 FCC's Request for Adopting RMA Standards in 1939

For facilitation of commercialization of television industry and pressure from industry for commercialization, FCC framed a rule for limited commercialization [214] and hence FCC requested all major television broadcasters in New York, Chicago, Los Angeles and Schenectady to adopt RMA standards. By the end of *May 1939*, various television set models covering size ranges from 5 to 14 in. sizes were made available by various companies like Andrea, DuMont and RCA in various departmental stores such as Macy's in New York. However, the companies offering better resolution specifications like Philco (Resolution od 605 lines and 24 frames per second) and DuMont (Resolution od 625 lines and 15 frames per second) did not agree to FCC imposed RMA standards. The exemption was granted by FCC based on Philco's request to adopt their television systems without any changes.

7.4 Conclusion

It is concluded that this chapter deals with inventions-related developments in television systems starting with still image sensing using direct contact and non-contact methods, image signal transmission and image reception resulting in developments of facsimile or fax machines. After the success of still image transmission, efforts were directed towards the transmission of moving images leading to the developments of television. The detailed developments of mechanical and all electronic developments are described in detail in this chapter.

Chapter 8
Efforts Directed Towards Vehicular Communication

Abstract The people felt the need of communication from vehicles and the effort of carrying radio equipment on moving platform like steam engine started as early as 1901. This long wave radio equipment was based on spark gap transmitter of Marconi Telegraph Company. Also during World War I, similar radio equipment was carried by Wagons with large number of storage batteries. In 1907, Alfred Grebe operated a crystal-based radio-telephone on car from long Island, N.Y. which was later on made using vacuum tubes in 1919, the high voltage required for vacuum tubes was generated using dynamotor. In 1908, Professor Albert Jahnke and The Oakland Transcontinental Aerial Telephone and Power Company claimed to have developed a wireless telephone, however, could not take efforts to produce it. In 1910, Frederick Baldwin and John McCurdy attempted for wireless radio communication between two planes, while in 1918, German Railroad System tested wireless telephony for their military trains. In 1921, base station in central Detroit was erected for the purpose of vehicular communication, however, as per the study by Robert Batts, power lines near it were compromising its operation and hence was moved to Belle Island in Detroit. Robert Batts modified radio equipment to use crystal control that was made to communicate with a vehicle fitted with wireless radio equipment having antenna running across the rooftop of the car. William P. Rutledge of Detroit police tried vehicular radio communication with the help of Robert Batts which was finally operational in 1928. Finally, in 1922, Gregorie Frost was successful in the integration of a radio equipment into a Ford model-T.

8.1 Efforts Directed Towards Building Portable Wireless Systems

At this point, mobile communication was very much desired due it inherent advantages and was free from wires which were required previously to transmit signals from sending stations to receiving stations and was very convenient. Mobile wireless radio equipment not just found the applications in safety but also found its way to traffic, taxi drivers and package deliveries. The mobile wireless has become key tool

© The Author(s), under exclusive license to Springer Nature Singapore Pte Ltd. 2021   147
V. Patil, *Chronological Developments of Wireless Radio Systems before World War II*,
https://doi.org/10.1007/978-981-33-4905-6_8

Fig. 8.1 The complete view
of Alfred Grebe's car with
antenna Construction of auto
radio telephone of 1919,
Credit A.H. Grebe, The Auto
Radiophone, Radio Amateur
News, August 1919, from
the articles and extracts
about early radio and related
technologies, concentrating
on the United States in the
period from 1897 to 1927,
United States Early Radio
History

in business leading to its rapid spread demanding more and mode frequency spectrum
and it had become a challenge to ensure efficient effective use of the spectrum.

8.1.1 Initiatives for Vehicular Communication

It all started with carrying radio equipments on moving platform, and these efforts
were stated as early as in *1901* when longwave wireless mobile station based on
spark gap transmitter of Marconi Telegraph Company was installed on steam engine
that covered the range of 30 miles. It had cylindrical antenna with large diameter
mounted on the rooftop of steam engine. In World War I from *1914* to *1918*, similar
radio equipment was carried by Wagons with large number of storage batteries being
carried along.

In *1907*, Alfred Grebe operated a crystal-based radio-telephone on car (Fig. 8.1)
from long Island, N.Y. which was later on made using vacuum tubes and in *1919*,
he constructed tuned frequency wireless radio receiver telephone working in 150 m
wave (2 MHz) amateur radio band that was installed in his car. The high voltage to
vacuum tube was applied using dynamotor and filament voltage was supplied by
battery, while antenna was constructed using six parallel wires.

In *1908*, Professor Albert Jahnke and The Oakland Transcontinental Aerial Telephone and Power Company claimed to have developed a wireless telephone and
could not take efforts to produce.

In *1910*, Frederick Baldwin and John McCurdy trailed behind another plane to
demonstrate the use of radio communication between two planes using an aerial
behind their plane were first to trail an antenna behind their bi-plane to demonstrate
the use of radio in aviation circles. In the beginning of year *1918*, German Railroad
System tested wireless telephony for their military trains in Berlin and a city in the
south of Berlin called Zossen.

Fig. 8.2 Detroit police car
fitted with mobile
equipment. The copyright
information was not found
and the photo is reproduced
without permission

The study of base radio station in central Detroit that was erected in *1921* was undertaken for its suitability for vehicular communication by Robert Batts and he concluded that its location was not suitable, the power lines near base radio station were compromising the operation of base radio station. The location of this station was moved to Belle Island in Detroit that was built as per the direction of Robert Batts. Robert Batts also modified to use crystal control. It was made to communicate with a vehicle fitted with wireless radio equipment having antenna running across the rooftop of the car [215, 216] as shown in Fig. 8.2.

In *1921*, William P. Rutledge [217, 218], Detroit police commissioner and public safety officer, was the first person to help the development of police communication by making an unsuccessful trial and error experiment to use radio for police work by purchasing Western Electric made 500 W radio transmitter that was installed in the central part of Detroit and also tried to build receiver to work on moving police cars but the noise was predominant and hence it was discontinued in *1926* due to its unsatisfactory performance. The main problem was the vehicle's heavy acceleration, sudden stops and electrical ignition noise were unbearable for the fragile radio receiver.

In *1927*, Robert Batts, an engineering student at Purdue University, who was working for his summer job in a "Detroit Radio Parts Store" known for providing the assistance for constructing radios from kits to its customers was of great help to this Detroit store. Kenneth Cox, a Detroit motorcycle patrolman, as one of the customers of Robert Batts had a discussion about how to make radio receivers work in moving police cars. Both continued their friendship even after Robert Batts joined back to Purdue University after finishing his summer assignment at "Detroit Radio Parts Store". Kenneth Cox continued to mail his designs to Robert Batts for corrections and rectification to make it work and one day he met William P. Rutledge to convince him to make an another attempt. The reliability of an equipment was demonstrated by dropping it on the floor that earned the approval of William P. Rutledge. It was finally operational in *1928*.

Gregorie Frost of Chicago at his 18 years of age integrated a radio into Ford model-T in *1922*. It became commercially available in *1927* and by the year *1934*, as many as 5000 land mobile radio equipped police cars and the age of mobile radio began.

The public trials of telephone started on trains between Berlin and Hamburg in *1924* while a company called Zugtelephonie A. G. was founded in *1925* to produce this telephone equipment while during *1926*, the telephone was made available in a train running on Berlin-Hamburg line for the first time and the wires running parallel to rail tracks worked as antenna.

On *September 25, 1928*, Paul V. Galvin and his brother, Joseph, founded a company called Galvin Manufacturing Corporation in Chicago, Illinois, USA, [219] became forerunner of Motorola, introduced first "Motorola Brand Car Radio" and this "Motorola Brand" meant "Sound in Motion" in *1930* and then Galvin Manufacturing Corporation started selling "Motorola" brand of car radios to public safely officers of police departments and municipalities. In *1936*, Galvin Manufacturing Corporation introduced the Motorola Police Cruiser one way radio receiver designed to receive police broadcasts tuned to single frequency specified by the customer.

8.1.2 Evolution of Handheld Transceivers

Although two-way radio communications were used in the existing years, it's need was not so intensive till World War II broke when need of handheld equipment was felt and efforts were directed in this direction during *1934* to *1941* and became popular as "Walkie-Talkie".

8.1.2.1 Donald Lewes Hings's Contributions

In *1930*, Canadian inventor Donald Lewes Hings developed portable radios for geophysical applications [220] for a mining company called Consolidated Mining & Smelting Company of Canada Limited became "Cominco" of Trail, British Columbia, Canada later on. He was challenged by a company head to develop a real-time communication between mines, aircrafts and various western and northern locations in Canada. During *1930–1938* he worked on lightweight aircraft radios, and as reported in an article on *June 07, 1930* under heading "Hings Broadcasts from Plane, Code Picked up Nelson", he developed lightweight aircraft radios using continental code (not voice) were being used by aircraft companies.

In *1937*, Donald Lewes Hings developed two-way emergency voice radio communication (C-58 production model) having range of 130 miles weighing 12 pounds with foldable antenna called "Light Aircraft Emergency Set". It used two frequencies one for "Cominco Company" and other for "Canadian Army Signal Corps" and was portable and can be carried while transmitting and receiving the voice signals that became a precursor to the "Walkie-Talkie".

During *December 07, 1938*, using the same technology, he developed "10PC20 Airplane Radio" similar to C-58 in size for "Cominco" Pilots and using same technology, Light Aircraft Emergency Set was made to communicate with small aircraft at longer distances that was capable of transmission of voice and code. It used two frequencies one was locked to Canadian Army Signal Corps while other frequency was locked to Consolidated Mining & Smelting Company of Canada Limited ("Cominco"). His special modulation scheme was called "Ever Expanding Modulation" that gave 20db gain on voice. In *1939*, Donald Lewes Hings approached his mining company Consolidated Mining & Smelting Company of Canada Limited ("Cominco") for taking the patent of his invention of two-way portable voice radios, however, Company was not interested in patents for communications. In search of his desire for patent, Hings travelled to Spokane, Washington across the British Columbia (BC) border, a city closer with an authorised Patent Attorney. A day after detailing his invention to Patent Attorney, on *September 10, 1939*, Canada had just declared war.

In *1938* Alfred J. Gross , a radio engineer succeeded in inventing a two-way handheld radio using miniatursed components. He explored frequencies above 100 MHz and built two 300 MHz models to transmit voice signal over 30 mils. He also contributed to Joan-Eleanor system developed in *1942* for US Office Strategic Services (forerunner of modern Central Intelligence Agency (CIA)).

During World War II, in *1940*, "Packset" was developed by Donald Lewes Hings based on small mobile emergency radio design that was used by "Cominco" and was demonstrated to Department of National Defence (DND) (Canada) and was considered to be too fragile that resulted in bringing out a tougher C-17.

In *1941*, Donald Lewes Hings says that two sets C 18 were taken to Toronto where soldier walked with it asked about "what does it do?", replied "Well you can talk with it, while you walk with it" and the name "Walkie Talkie" came in existence and the first use of "Walkie Talkie" made in *1941*.

8.1.2.2 Galvin Manufacturing Corporation's Contributions

In *1939* Galvin Manufacturing Corporation (Forerunner of Motorola) founded by Paul V Galvin in *September 25, 1928*, designed and developed two-way voice communication radio "Handy-Talkie" also called "Fightingest" radio set used in front lines of army employing miniature glass tubes with low filament voltages. Earlier, it was basically developed for portable civil applications (for public safety officers moving on the cars) in the form of 1R5, 1S5 and 1T4 announced in *1939*, used superheterodyne receiver circuit with local oscillator and RF stage converted to master oscillator-power amplifier for transmission. It was basically a crystal controlled AM receiver operating on 3.5 and 6 MHz HF frequencies.

8.2 Conclusion

The efforts were made by Alfred Grebe, Albert Jahnke & The Oakland Transcontinental Aerial Telephone and Power Company, Frederick Baldwin & John McCurdy, William P. Rutledge and Gregorie Frost for making radio equipment suitable for vehicular communication. Donald Lewes Hings developed portable radios geophysical applications that became precursor for development of "Walkie Talkie".

Chapter 9
Status of Wireless Air Defence Systems Before Word War II

Abstract It is all about the status of wireless radio systems before the start of World War II when development and deployment of defence systems happened at a greater pace. Actually, these were the actions in response to German aggression and defence against German U-Boats and aircrafts and the havoc they played. The enemy targets either moving or stationary were under the scanner. It was important to locate them and destroy and hence how to locate them was the subject of the interest and almost all targets were using radio waves for either communication or navigation purpose and hence it was a natural choice to observe from which direction radio waves are coming from. This led to the development of radio direction finders. Bellini-Tosi, Adcock contributed to these initial developments and initial efforts were to detect static targets which were later on extended to moving targets by Robert Watson-Watt who designed Chain Home radio direction finders, and later these techniques were called radio detection and ranging (RADAR). The Chain Home direction finders involved both static transmitting and receiving stations at one place giving birth to mono-static radar concept and further to get better accuracy of detection, bi-static and multi-static concepts were evolved. These earlier detection mechanisms required huge and heavy static infrastructure and also requirement of uninterrupted function during night, and also need for airborne systems was felt. Robert Watson-Watt split his group into two working groups under Wilkins and Bowen, respectively. The continuation of work on chain home radars was entrusted to Wilkins and while work on night interception radar was assigned to Dr Edward George "Taffy" Bowen who worked on airborne radar systems and successfully developed it.

9.1 Effects of World Wars on Wireless Systems Developments

The wide-ranging wireless systems such as broadcast radio systems, television systems, target detection systems in the form of direction finders, coding of information for security purposes had been existed even before World War II. It was a World War II that provided accelerated thought process about how to use these systems for various military purposes and sparked the growth of wide-ranging wireless systems for tactical warfare and defence applications. The use of wireless radio communication equipment was well established even before World War I.

9.1.1 Role of U-Boats During World Wars

The U-Boat stands for "undersea boat" and stands for German U-Boot an abbreviation for "unterseeboot" was an effective weapon against allied enemy forces. There were 38 U-Boats at the beginning of World War I which were increased to 334 at the end of World War I. However, only 60 were active at a time with the exception of 140 U-Boats active in *October 1917*. To combat German U-Boats, Britain developed antisubmarine submarines which were responsible for sinking of 17 German U-Boats during World War I. U-Boats becomes subject of detection of targets that was evolved with the passage of time. Since targets were always equipped with radio equipments for communication, initial efforts were to just find or detect the source of radio frequency waves and to find from which direction these radio waves are coming from to track down the target and such systems were called "radio direction finders (RDF or DF)".

9.1.2 Radio Direction Finding

U-Boats becomes subject of detection of targets that was evolved with the passage of time. Since targets were always equipped with radio equipments for communication, initial efforts were to just find the source of radio waves and from which direction these radio waves are coming to track down the radioactive target and such systems called "radio direction finders (RDF or DF)" were mostly used for finding locations of radio frequency sources of enemy radio equipment installations, finding locations of non-authorised transmitters, searching of radio frequency interference sources, radio communication sources of criminal organisations, military intelligence for gaining the information about enemy communication or spying during war time and for research purposes. Besides these applications, the direction finders are also of great use for navigation purposes for ships and aircrafts. The systems used for the detection of radio active targets were receive only systems.

It all started with the detection of radioactive targets but what about passive target that does not emit radio signals? In such cases, the option was to illuminate the target by flooding radio frequency radiation and receive back the reflected radio frequency signals to detect the target that used transmit and receive systems and at this point, terminology typically called radio detection and ranging (RADAR or simply called Radar) came in to the existence. Fig. 9.1 shows how radio and non-radioactive targets are detected.

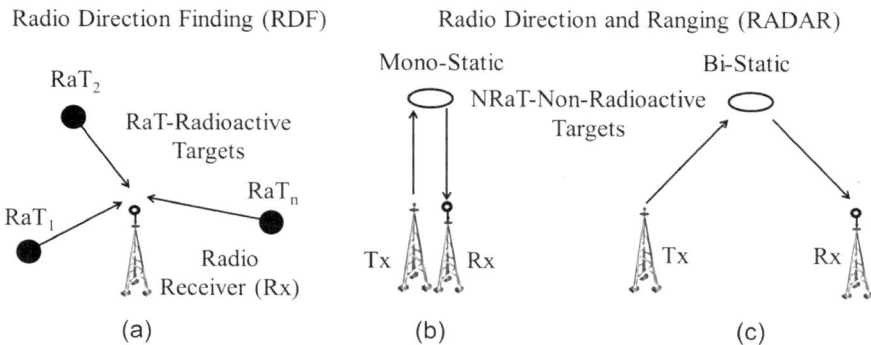

Fig. 9.1 Radioactive and non-radioactive target detection. **a** Radio direction finding (RDF). **b** Monostatic radar and **c** Bi-static radar

Fig. 9.1a shows the method of radioactive target detection called radio direction finding (RDF) while passive targets are detected by two methods, in the first method radio transmitter and receiver are collocated or located at one place called monostatic radar (Fig. 9.1b) but when radio transmitter and receiver are located reasonably at longer distances it is called bi-static radar (Fig. 9.1c) and these both methods described in Fig. 9.1b, c are widely classified under radio detection and ranging (RADAR or Radar).

9.1.3 Use of Direction Finding for Navigational Purposes

Till now we focused our attention on the direction finding of enemy targets only; however, there were also some friendly aspects of radio direction to use it for the purpose of navigation. It was not that methods of navigation did not exist but traditional ways were also in existence. The radio direction finding (RDF) opened a new era for application of direction finding not only for enemy target detection but also for navigational purposes.

9.1.3.1 Traditional Ways of Direction Finding

The traditional ways for direction finding had been widely used for the purpose of aircraft and ship navigation from ancient times. In absence of radio communication in the early days, position of ships or aircrafts was determined either by dead reckoning (DR) or by Astronomical observations [221]. The dead reckoning (DR) is the method of tracing the path using compass observations, time of travel and speed of an aircraft or ships. In DR navigation, the inaccuracies were added due to wind speed and aircraft speed especially when the aircraft frequently changes its course of direction. The Astro observations were typically applied to the long flights and dependent on weather conditions. Although older methods were acceptable for ship navigation to

Fig. 9.2 Principle of radio direction finding

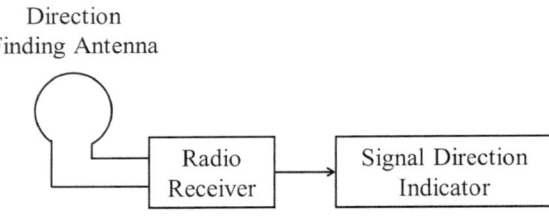

Direction Finding Antenna

Radio Receiver

Signal Direction Indicator

some extent, posed great difficulties for aircraft navigation due to its speed and hence radio direction finders took first place in navigation of aircrafts.

9.1.4 Principle of Radio Direction Finding

The early radio direction finders were based on simply the detection of radio signals emitted by enemy radio installations or enemy ships or enemy aircrafts or enemy mobile installations. The radio detection requires directional antenna, radio receiver and radio signal indication. Antenna detects the direction of the incoming signal and its position becomes indication. The position of antenna where maximum or minimum signal received (depending on antenna arrangement) becomes detection point. The principle of direction finding can be explained using Fig. 9.2.

From above circumstances and descriptions, it is apparent that radio frequency detection occupied first place in the priority of target detection and scientific community focused their attention to invent the various ways of target identification or detection using radio frequencies.

9.2 Early Efforts and Inventions in Radio Frequency Target Detection

The target detection relied on the sensing of radio frequency waves emitted by the target or sensing radio frequency waves reflected by the enemy targets. In the process, the direction from which radio waves are being received becomes the basis for detection of the target location and such systems can also be called radio frequency direction finding systems.

9.2.1 Heinrich Hertz's Experiments

In Heinrich Hertz's experiments [222], on _December 1887_, showed that electromagnetic air waves were shielded and reflected by metallic plates or surfaces that became a landmark invention for target detection.

9.2.2 Alexander Stepanovich Popov's Work

In USSR during *1895*, Alexander Stepanovich Popov [223, 224], working at Imperial Russian Navy school in Kronstadt, worked on coherer tube for detection of distant lightening strikes and added a spark-gap transmitter to make it as a complete system for communication between two ships in Baltic region. In *1897*, while communication with other ship, third ship passing by caused interference beat, made him to conclude that this phenomenon can be used for detecting the objects.

9.2.3 Guglielmo Marconi's Experiments

On *March 3, 1899*, Guglielmo Marconi, while conducting the radio beacon experiments at Salisbury Plain, noticed that signal transmitted by the transmitter was reflected back to the transmitter, however, he could not anticipate the use of such phenomenon for a particular application and rediscovered again in *1922* in his paper before the Institution of Electrical Engineers in London.

9.2.4 John Stone Stone's Experiment to Determine Direction of Space Telegraph Signals

In *1902*, John Stone Stone [225] worked on determination of direction of space telegraph signals, used two parallel conductors and loop type configurations of the antennas which is described in his US patent No.716134 entitled "Method of determining the direction of space telegraph signals" dated *December 16, 1902*. In this method, position of the conductors or the part of conductors is so moved relative to the direction of motion of electromagnetic waves till position is found.

9.2.5 Christian Hülsmeyer's Demonstration

Meanwhile, on *May 17, 1904*, Christian Hülsmeyer, held the public demonstration [226] of his equipment called "Telemobiloskope" [227, 228] at the yard of the Dom-Hotel in Cologne in front of shipping companies and immediately on the next day at Hohenzollern Bridge (in Cologne) across the river Rhine. It used spark gap transmitter producing radio waves of wavelengths between 40 cm (749 MHz) and 50 cm (600 MHz) located on high tower while receiver was located on Hohenzollern Bridge. The experiment showed that radio echoes or reflected radio waves can be used for ship detection and navigation purpose. His experiment consisted of transmission con-

Fig. 9.3   The Hülsmeyer's first radar device called "Telemobiloscope" in 1904   *Source* Deutsches Schifffahrts Museum, Leibniz-Institut für Maritime Geschichte, Hans-Scharoun-Platz 1 D-27568 Bremerhaven, Germany

sisting dipole antenna with parabolic reflector and similar type of coherer receiver for detection of ship.

On *April 30, 1904* he filed a German patent No. DE165546 entitled "Verfahren, um entfernte metallische Gegenstände mittels elektrischer Wellen einem Beobachter zu melden" and English patent entitled "Hertzian-wave Projecting and Receiving Apparatus Adapted to Indicate or Give Warning of the Presence of a Metallic Body, such as a Ship or a Train, in the Line of Projection of such Waves" on *June 10, 1904*. His invention provided a basis for "Radio Deatection and Ranging". The Christian Hülsmeyer's mono-static "Telemobiloskope" (Fig. 9.3) was first rejected by none other than Admiral von Tirpitz of the German Navy and made comment that "Not interested, my people have better ideas".

9.2.6   Otto Scheller's Homing System

Locating the target and moving towards it, it is a form of directional radio which is also used as a device for guiding the pilots to reach the airports or home and hence called radio homing system that made navigation practically possible in cloudy environment. In *1906*, Otto Scheller patented a homing system [229] a kind of DF system that used two antennas pointing towards slightly different directions while pilot was provided with only simple radio receiver to provide beep when aircraft follows intended airway.

9.2.7   Bellini and Tosi's DF System

In *1907*, Ettore Bellini and Alessandro Tosi [230] used two static perpendicular loop antennas instead of rotating antenna loop for direction finding that greatly improved the design of the radio direction finding. The perpendicular arrangement of loop antennas provided the help in finding sine and cosine components to determine the direction of the radio signal.

9.2.8 Robert Watson-Watt's Work on Detection of Lightening and Thunderstorms

In *1916*, Robert Watson-Watt worked at Air Ministry Meteorological Office at Aldershot, Hampshire, UK and was interested in use of radio signals for detecting thunderstorms and lightning and worked on system to accurately track the approaching "Lightening and Thunder Storms" [231, 232] for warning the pilots. In fact lightening was also a major problem for wireless communication at common wavelengths. His early experiments using rotating antenna loop could detect the signals from the range up to 2500 kms. which became a precursor for developments of direction finding equipments, however, its role in military intelligence was not established before *1930s*. In the same year, he proposed the use of cathode-ray tube (CRT) as an indicating device instead of mechanical indicating device.

9.2.9 Frank Adcock's DF System

The loop-based system worked well for vertically polarised signal component but failed miserably to work for horizontally polarized components which were significant during night time due to behaviour of ionosphere. This problem was solved by British engineer Frank Adcock [233] in *1919* using an antenna called Adcock Antenna consisting of monopole or dipole antennas sensitive to the direction of radio signals.

9.2.10 Guglielmo Marconi's Afterthought

In *1922*, Guglielmo Marconi recognized the possibility of detecting metallic objects using reflected radio waves and his thought was picked up by two American inventors called Albert H. Taylor and Leo C Yong.

9.2.11 Albert Hoyt Taylor and Leo C Yong's Moving Target Detection Experiment

In Mid-September *1922*, Albert Hoyt. Taylor and Leo Clifford Young while working at Naval Aircraft Radio Laboratory, Anacostia, DC, USA picked up Guglielmo Marconi's thought of metallic object detection and started experimenting it on the moving targets using superheterodyne receiver and 50 W amplitude-modulated transmitter using 60 MHz continuous wave (CW) radio signal modulated 500 Hz signal and noticed that when electromagnetic beam directed towards moving target (a wooden ship in Potomac river) [234], they received back the strong signal due to interference of transmitted signal and signal reflected from the target. This was the first continuous wave (CW) interference detector demonstration that also served as the first unambiguous evidence of bi-static principle of radio detection. Albert Hoyt. Taylor and Leo Clifford Young thought that this phenomenon could be used for detecting intruding enemy ships in the night. In *September 27, 1922*, they wrote a letter to Navy Bureau of Engineering [235] suggesting that this phenomenon can also be used to trigger an alarm to detect moving objects in space. However, they failed to take immediate follow-up on the issue and could not encash on their idea on moving target detection either on surface or in the air space.

9.2.12 Establishment of NRL

The prominent radio laboratories like Naval Radio Telegraphic Laboratory (established in *1908*) and the Naval Aircraft Radio Laboratory (established in *1918*) were already in existence, and consolidation of these laboratories was happened in the form of Naval Research Laboratory (NRL) [236] on *April 16, 1923* while it was officially commissioned on *July 02, 1923* and Albert Hoyt Taylor was made as first Superintendent of this laboratory.

9.2.13 H. M. Mottez's Work on Antenna Directivity

By the end of World War I rotating directional antenna loops versions were employed in direction finders, however, these loops were inefficient and range was short and this problem was solved in *April 24, 1923* by H.M. Mottez [237] (French Patent No. 577629 dated *April 24, 1923*) who employed high efficiency non-rotating large vertical antennas and to obtain rotating signal pattern he employed varying the phase of the signals fed to them. Although the technic was not new and was being used for beaming the transmission by MW transmitters, it provided efficiency and range improvements.

9.2.14 Appleton and Barnett Use of FM Radar for Detection of Ionosphere

In UK, E.V. Appleton and M.A.F. Barnett [238] used FM radar to measure the height of ionosphere in *December 1924* and in the following year the similar measurements on ionosphere were done by Breit and Tuve in USA using pulsed radar technique.

9.2.15 Gregory Breit and Merle Anthony Tuve's Contribution

In *1925*, yet another pair of Americans called Gregory Breit and Merle Anthony Tuve [239] working at Department of Terrestrial Magnetism (DTM), Carnegie Institution of Washington, Washington DC, were the first to use broadband pulses ranging from 3 to 30 MHz to measure the heights of the different layers of ionosphere and were regarded as the first to successfully use the pulse technique in detection which was then widely used in later radio detection and ranging.

9.2.16 Pierre David's Electromagnetic Barrier

In *June 1925*, French pioneer Pierre David [235] working with Laboratoire Central de TSF (Central Laboratory of Wireless-Telegraphy) created after WW-I by General Gustave Ferrie the initiator of radio technology was first to take interest in problem of aircraft detection using metric waves, observed that waves were frequently interrupted due to electromagnetic noises generated by the spark plugs of nearby combustion engines that were used in motorcycles and cars. The matter was reported to General Gustave Ferrie who inturn suggested that by listening to the radio noise produced by ignition of aircraft ignitions, the aircrafts can be detected.

9.2.17 Herman A. Affel's Position Indicator

Herman A. Affel (*May 5, 1923*) [237] used two aerials spaced apart fed by single transmitter, emitted the signal at same frequency and phase to be used as "Movement and position indicator" and it was patented vide US Patent No. 1562485, *24 November 1925*. He also being regarded to have invented the practical coaxial cables which become highly useful in radar and microwave applications.

9.2.18 Estill I. Green's Range Finding Between Two Stations

On *December 8, 1927*, Estill I. Green applied for "United States Patent No.1750688 on *March 18, 1930* entitled "Determining Movement And Position of Mowing Objects" for range finding between two transmitting and receiving stations in which he employed modulation of signal on a certain frequency and sent it to another station and same signal was sent back to a original station on another frequency maintaining the phase of modulation.

9.2.19   Albersheim and Konheim's Work

Walter Albersheim (*January 25, 1929*) and Harvey Konheim (*January 25, 1929*) (US Patent No. 1995285, *March 26, 1929*) entitled "Radio Navigation" in which they dropped the phase maintenance logic but maintained the same amplitude of both but the different carriers are linked to multiplication factor to derive the common frequency. Slave was being phased locked to master signal that permitted longer baseline between two stations. This led to an idea of hyperbolic systems in which a flying object can be sensed by a pair of transmitting and receiving antennas. Many worked on similar lines as that of E.I. Green and Albersheim & Konheim which include a German inventor Meint Harms (*May 20, 1930*) who worked on harmonically linked frequencies(German Patent No. DE546000, *March 8, 1932*); E.A.H. Honore (*July 31, 1935*), (US Patent 2148267, *February 21, 1939*) entitled "Method of and Apparatus for Radio Goniometric Indication") successfully worked on accurate phase-locking between two transmitters while John P. Shanklin (*September 23, 1937*), (US Patent 2144203, *January 17, 1939*) entitled "Method for Direct Indication of Position in a Given Area") worked on multiple harmonically-related modulation frequencies.

9.2.20   Accidental Spotting of Aircraft by Young and Hyland

In *June 24, 1930*, Leo C Yong and Lawrence A. Hyland [240] of Naval Research Laboratory of USA, while working on 30MHz radio compass using an directional property antenna working on 32.8 MHz, accidently spotted the disturbance when aircraft was passing from a distance of 3.2 kms. Leo C Yong and L.A. Hyland interpreted this phenomenon as the "Effect of Beats" or "Radio Interference Effect". This was the first instance of location of aircraft using radio waves.

9.2.21   Origination of Bi-static Concept

In *June 03, 1932*, British Post Office engineers, Nancarrow F.E., Mumford A.C., Carter P.C., and H.T. Mitchell observed the bi-static [241] directions and it was also became one of the landmarks in futuristic bi-static radio detection methods.

9.2.22   Taylor, Young and Hyland's Use of Bi-static Radar

In On *June 13, 1933*, Taylor, Young and Hyland [240, 242] experimented bi-static principle of radar for the detection of moving target using transmitter and receiver located at the distance of 5.5 kms separated by hilly areas that allowed the detection of movement of aircraft at the distance of 50 kms. The disclosure in the form of patent was made in 1934.

9.2.23   Rudolf Kühnhold's Efforts

In *1933*, Rudolf Kühnhold unsuccessfully attempted to use microwaves at 2.22 GHz (13.5 cm wavelength) for detection purpose in which microwaves were produced using first ever produced microwave tube generator called Barkhausen–Kurz tube, asked help from amateurists Paul-Günther Erbslöh and Hans-Karl Freiherr von Willisen working on VHF communication system which they agreed and in January 1934 they set up a company called Gesellschaft für Elektroakustische und Mechanische Apparate (GEMA).

GEMA hired two consultants Hans Erich Hollmann and Theodor Schultes from Heinrich Hertz Institute in Berlin to work on radio measuring and detection equipment that used magnetron purchased from Philips Netherland. Hans Hollmann worked on regenerative receiver while Theodor Schultes concentrated on the developments of Yagi antennas for transmission and reception. During *June 1934*, a vessel at the distance of 2 kms passing trough Kiel harbour was detected using Doppler interference.

Afterwards, GEMA started working with pulse modulated system for better frequency stability of Philips Magnetron working at 600 MHz (50 cm wavelength). The pulses of 2 μs with a repetition rate 2000 Hz were used. The transmitting antenna consisted of 10 pairs dipole antennas with a background reflecting mesh and out of ten 10 pairs dipole antennas, three pairs are used for receiving. The duplexer device is used to shut the receiver when transmitter is pulsed. The resultant range is displayed on cathode ray tube (CRT) (Braun tube) and could detect ships at the distance of 10 kms.

Using similar concepts the fist successful radar was built in *1935* for ship born applications operated on 80 cm. The pulsed radars [243] were distinguished as land-based radar working on VHF frequency range 120–130 MHz called "Freya" while sea-based radar working on UHF frequencies 600, 500 and 390 MHz was called "Seetakt" which were operative by 1940.

9.2.24   French E'mile Girardeau's Radar

In *1934*, E'mile Girardeau started working towards development of rudimentary radar for installation on ships and land [243] based on principles stated by Nicola Tesla and came out with first French radar. He obtained a French Patent No 788795 in *1934*. In the next year, his radar device installed onboard of cargo ship Oregon and Normandie during next year, that is *1935*.

9.2.25 Robert Morris Page and Leo C. Young's Contributions

Further, the Director of US Naval Research Laboratory(NRL), Robert Morris Page was then assigned a work of development of 60MHz pulsed transmitter by Albert H. Taylor to get the pulse of duration of $10\,\mu s$ with the spacing duration of $90\,\mu s$ between them. The transmitter was then used for the detection of airplane flying over Potomac river at the distance of 1 mile in *December 1934*. Robert Morris Page and Leo C. Young are usually credited with true radar development.

9.2.26 Watson Watt's Aircraft Detection Experiment at Daventry

On *February 26, 1935*, Watson Watt's conducted an experiment to aircraft called Daventry experiment, successfully tracked the aircraft on cathode-ray tube oscilloscope which was later developed in the form of chain home radar.

9.2.27 Invention of Duplexer

In *1936*, the invention of duplexer facilitated issue of isolation of transmitter and receiver at single site useful for aircrafts, ships and mobile ground units.

9.3 Detailed Description of Radio Direction Finding

The radio direction finding properties can be vested either in radio transmitting or receiving equipment. The two types of options that can be implemented are option-1: Fixed omnidirectional radio transmitter and direction finding radio receiving system on mobile station like aircraft or ship while other implementation can have option2: Omnidirectional radio transmitter on mobile station and fixed direction finding radio receiving system on shore.

9.3.1 Overview of Radio Direction Finding Methods

The antennas are mainly used for transmitting and receiving radio signals, however, when antenna is used in receiving mode for identifying direction of incoming radio signal, one can locate the object that is emitting the radio frequency signals becomes an important tool for radio frequency direction finding and detection enemy targets.

Fig. 9.4 Direction finding
coil

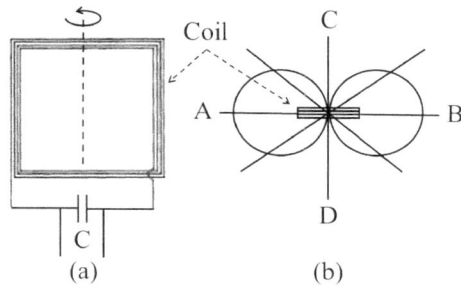

(a)   (b)

9.3.1.1   Loop Antenna Direction Finder

The radio direction finding was extensively useful technique for both naval and
aerial navigation even before World War I using single circular loop antenna whose
circumference was being decided by the frequency of the signal to be detected. If the
loop was being placed in the direction perpendicular to the incoming signal, then the
signal intensity is zero. The signal in two halves cancel each other to get null point
that is responsible for signal detection and in order to get signal source detected,
loop was mechanically rotated till signal disappears means null is detected [244].
Principle of direction finding is shown in Fig. 9.4 where Fig. 9.4a shows multiple
turn coil while Fig. 9.4b shows how received signal intensity varies, demonstrating
the directional properties of the loop coil antenna.

The principle of direction finding explained above requires loop antenna to be
rotated to find the direction from where signal is coming and to eliminate the moving
parts in signal sensing, some of the inventors like Bellini and Tosi thought of two
static perpendicular coils to find the direction of incoming radio signal. The two
signal components sensed by two coils representing sine and cosine components
were then conveniently used to find the direction of the signal.

9.3.1.2   Bellini–Tosi Direction Finder

Ettore Bellini and Alessandro Tosi greatly simplified the direction finding equipment
and also replaced single loop antenna by two perpendicular field antenna loops. This
arrangement removed the need for mechanical movements of the antenna system.

The direction finding was one of the most important inventions of Ettore Bellini
and Alessandro Tosi, in _1907_ [245] also called radiogoniometer that determines
angular direction of incoming radio signal. This equipment picks up maximum field
strength from specified source signal direction and hence also called Bellini–Tosi
direction finder(B-T DF or BTDF). The basic working principle of Bellini–Tosi
direction finding equipment is shown in Fig. 9.5.

The equipment consists of two loop antennae mounted along the planes at 90°
each other extracting sine and cosine components of the signals of target angle to find

Fig. 9.5 Basic principle of Bellini–Tosi direction finding

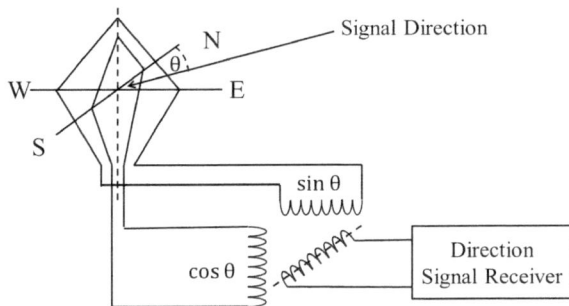

Fig. 9.6 Photographs of Royal Navy signals establishment's S25 model of radiogoniometer used in Bellini–Tosi direction finder, Courtesy: HMS Collingwood Heritage Collection, Signal School's Photograph No. SS689

out the position or direction of the target. One loop is perpendicular to north-south direction while other loop is perpendicular to east-west direction. The photograph of coils which were connected to antenna loops is shown in Fig. 9.6 radiogoniometer

It was patented vide US Patent No. 943960 entitled "System of Directed Wireless Telegraphy" dated *December 21, 1909*, the application for which was filed on *October 01, 1907* [230]. Some times it was also called Marconi-Bellani-Tosi direction finder after they sold their patent rights to Marconi Company in *February, 1912* and joined Marconi Company. The use of direction finders was also strongly advocated for navigation [244] by the Department of Commerce on the Interdepartmental Radio Committee on Safety at Sea on *May 1913*. The use of radio communication equipment was well at the time of outbreak of World War I in *1914*, and then it was to determine the relative bearing of low frequency transmitters and could not provide significant results even after study made by HMS signal school.

The early and first successful use of Bellini–Tosi direction finders for on board navigation purpose was during *1916*. These systems were also used to detect enemy wireless positions by British Army and Royal Navy for detection of German submarines. Henry J. Round working with military intelligence in the rank of captain set up a chain of direction finding stations in England which started monitoring transmissions and proved to be useful in detecting signals from German navy ships anchored at Wilhelmshaven [246] also later on in *May 30 1916*, the change in position of German fleets was also detected. Also Henry.J. Round's innovation of valve-based direction finding amplifier offered great transformation in military intelligence. Early direc-

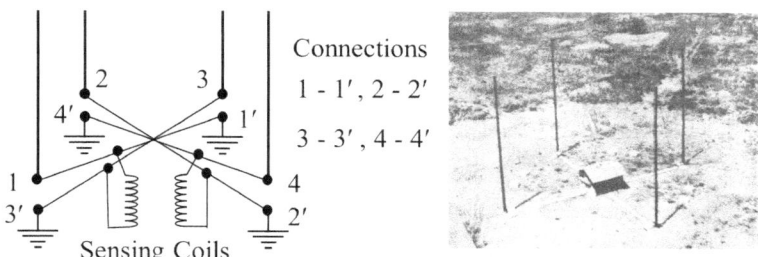

Fig. 9.7 Adcock antenna constructional details and installation at Rabaul in East New Britain province, Credit: CC BY-SA 3.0, P.I.C., Anacostia, D. C, | January 1, 1945| 90-foot (27 m) diagonal spacing Japanese Adcock direction finder installation for 2 MHz in Rabaul

tion finders used medium and low frequencies which were not suitable for highly efficient direction finder antenna.

The success of Bellini–Tosi direction finders was based on the skill with which antennae were designed [247]. During *1920* and *1930*, many types of antennas were tried which include two bridge structured coils suitable for large ships that were replaced by two large loops, two loops with one of rectangular shape in fore and aft plane and triangular shape at right angle or some angle. Both loops covered large area to sense long waves at any distance.

9.3.1.3 Frank Adcock's Antenna for Providing Immunity Against Effects of Sky Waves

The direction finder was an effective tool for enemy targets during World War I for lower frequency waves but failed at higher frequencies. At higher frequencies, many signals arriving from ionosphere due to reflections from multiple directions posed a serious problem in detection finding. The loop-based system worked well for vertically polarised signal component but failed miserably to work for horizontally polarized components which were significant during night time due to behaviour of ionosphere. This problem was solved by British engineer Frank Adcock in *1919* using an antenna called Adcock Antenna (British Patent No. 130490) [233] that was based on symmetrically interconnected pairs of vertical monopoles or four dipoles. The constructional details of Adcock antenna are shown in Fig. 9.7. The vertical elements of an antenna provided great immunity against the effects of horizontally polarized signals even presence of sky waves.

9.4   Robert Watson-Watt's Job at NPL

In *1924*, the occupation of Aldershot site by War Department, resulted in to the movement of Robert Watson-Watt to Ditton Park near Slough, Berkshire where National
Physical Laboratory (NPL) was already in existence. In *1927*, an amalgamation with
the National Physical Laboratory, Radio Research Station was established wherein
he became Superintendent of an NPL outstation.

9.5   Shift of Interest Towards High Frequency Direction
Finding (HF/DF)

German U-boats installed with high frequency equipments posed a serious threat to
allied forces and HF/DF direction finding was an effective technique against finding
accurate positioning of U-boats [248]. The ships installed with HF/DF systems can
find the source of high frequency signal used by U-boat and move towards such
signal source to destroy it. Locating the target and moving towards it, it is a form of
directional radio which is also used as a device for guiding the pilots to the airports or
home and hence called radio homing systems [249]. This term was also being used
with "Marconi Robinson homing systems". The great advantage of such direction
finding system is that it does not require two-way communication.

During *1930*, instead of two loops, single circular rotating loop was adopted
for installation on warships like HMS Barham. Afterwards, it was also realized
that for use with high frequencies, better alternative was to use smaller area double
loop instead of large area single loop that was only suitable for lower frequency.
The rotating diamond-shaped double loop antenna with smaller area was then more
suitable for installation on smaller destroyer ships for successful for high frequency
operation. After *1931*, much of the attention was paid towards the development of
high frequency direction finders also called HF/DF or HFDF or Huff Duff. The trails
for rotating antenna were conducted on HM Cruiser "Concord" by fitting it on the
top spot in the roof on the top of foremast, however, the results were not free from
errors due to ship's length and rigging.

9.6   Pre-world War II Developments

The war like situation leading to World War II was responsible to give birth for radar
concept in which three stakeholders like RF transmitter, RF receiver and target are
necessary.

9.6.1 Britain's Efforts Against German Aggression

The reorganisation in *1933*, Robert Watson-Watt became Superintendent of radio research department of NPL at Teddington where he started and concentrated on work of detection of aircraft using radio signals.

9.6.2 The Formation of Committee for Scientific Survey of Air Defence (CSSAD) by Britain

This was the time when the foundations for defence against air attack were also laid in *November, 1934* when a committee for Scientific Survey of air defence (CSSAD) to conduct scientific study [250] was formed for strengthening the defence against aircraft attacks was formed in Great Britain. The committee was headed by Henry Tizard, consisted of four members such as Henry Tizard, H.E. Wimperes and two scientists.

9.6.3 The First Meeting of CSSAD

It's first meeting of CSSAD was held in *January 1935* in London under the chairmanship of Henry Tizard, and in this meeting the notable personalities like Archibald Vivian Hill a physiologist, a Nobel Laureate and young professor of physics, Patrick Blackett and Albert Percival Rowe were present and became key players in laying the foundations for a story of radars and radio direction finders (RDF). The measures were suggested to harm the pilots of invading aircraft. At this point, role of Robert Watson-Watt [251] was very much needed due his background in radio detection. Following the first meeting of CSSAD, Arnold F. Wilkins suggested to Watson Watt that radio waves could be used for detection of enemy aircraft detection.

9.6.4 Involvements of Robert Watson-Watt by CSSAD

The CSSAD suggested to find out the methods for either to harm the invading aircraft or harm the pilot in terms raising his body temperature at least by $2\,^\circ$C using radio waves (Death Ray) and Robert Watson-Watt was asked to comment and write a two memoranda on the feasibilities such methodologies. Robert Watson-Watt gave an assignment to Arnold F. Wilkins, a scientific officer [251, 252] of NPL to calculate the requirement of radio frequency power to raise the temperature of the water of 8 pints (One Pint = 568.26 cubic centimetres) from 98°F to 105°F kept at the distance five kilometers and at the height of 1 kms. He ruled out the possibility of unconvincing the pilot of even low flying aircraft while on other front that of raising the body temperature of the pilot of invading aircraft, he suggested that to raise 2 °C temperature, it will require transmitter that can supply 5 Giga Watts power to transmitting dipole antenna. However, the application of this idea can re-radiate sufficient energy for detection purpose and tracking can be more lucrative rather than producing death ray for pilots.

9.6.5 Watson-Watt's Daventry Aircraft Detection Experiment: A Precursor of Radar Systems Developments

On *February 06, 1935* Robert Watson-Watt wrote a letter to A. P. Rowe, enclosing the memorandum on "Detection of Aircraft by Radio means" and by *February 12, 1935*, the official reactions were received on first memorandum submitted by Robert Watson-Watt and funds of £10,000 were made available by Air Marshall Sir Hugh Dowding for experiment. The test flight was arranged between 10 kW BBC short-wave transmitter with call sign GSA (operating on shortwave frequency of 6.13 MHz or 49.8 m wavelength) at Daventry [253] and town of Weedon on 26 February, *1935* while on ground, a small vehicle containing receiving equipment consisting of radio receiver and cathode ray tube oscillograph for viewing the result. The reflected RF signals from the aircraft were received using two dipole antennas that were connected to receiving equipment located in vehicle. The moving metal-winged Heyford bomber aircraft, number K6902 piloted by Flight Lieutenant R. S. Blucke of Royal Air Force (RAF), was tracked on oscilloscope trace and became a precursor of war time radar system developments and Watson-Watt and Wilkins's theory was proved. This experiment typically called Daventry experiment is shown in Fig. 9.8.

The "Daventry Experiment" used frequency of frequency of 6.13 MHz (49.8m wavelength) with logic of half wavelength matches with the wing span of enemy aircraft (Heinkel 111) having a wingspan of 22.5 m and this logic was abandoned later. The received power is governed by Eq. 9.1 as follows

$$P_r = \frac{P_t G_t G_r \lambda^2 \sigma}{(4\pi)^3 R_t^2 R_r^2} \tag{9.1}$$

where P_r is received power, P_t is received power, G_r is receiver gain, G_t is transmitter gain, λ is wavelength, σ radar correctional area, R_t and R_r are the distances of target from receiving and transmitting antennas. While range Eq. 9.2 is as follows

$$\kappa = (R_T R_R)_{max} = \left[\frac{P_T G_T G_R \lambda^2 \sigma_B F_T^2 F_R^2}{(4\pi)^3 k T_S B_n L_T L_R (S/N)_{min}} \right]^{\frac{1}{2}} \tag{9.2}$$

where κ is Range, R_T, R_R are transmitter to target and target to receiver ranges (m), P_T is transmitted power output (W), G_T, G_R are transmitting and receiving antenna

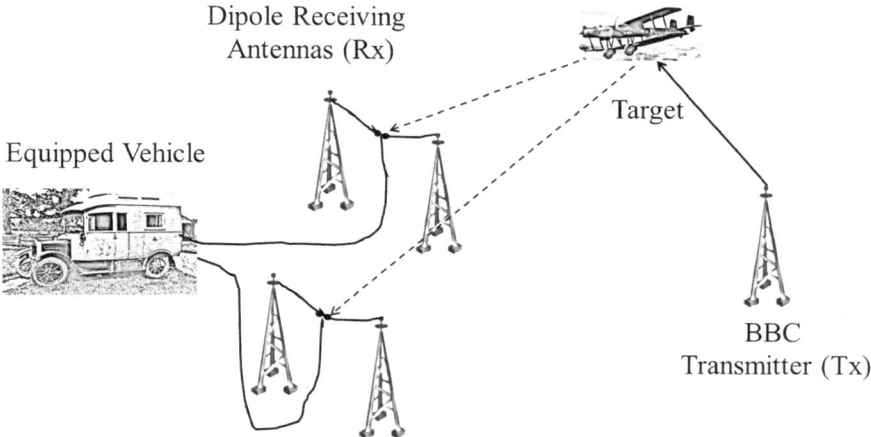

Fig. 9.8 Watson-Watt's Daventry aircraft detection experiment served as precursor of Radar developments

power gains, λ is transmitted frequency wavelength, σ_B target bi-static radar cross section (m^2), F_T, F_R transmitter to target and target to receiver pattern propagation factors, k Boltzmann's constant (J/K), T_S is receiver noise temperature (K), B_n receiver noise bandwidth(Hz), $(S/N)_{min}$ is minimum signal to noise ratio, L_T, L_R are transmitting and receiving system losses and κ is maximum bi-static range product.

9.7 Radio Detection and Ranging (Radar) Concept

In earlier cases of detection, radioactive targets were involved and radio emission from the target was exploited for detection purpose. The concept of active and passive radar came in existence when people started exploring the detection of non-radioactive targets for which dedicated radio source or dedicated transmitter is used to illuminate the target so that reflected radiation can be detected to locate such target and it became apparent to use a dedicated pair of transmitter and receiver for target detection to become non-reliant on the deliberate use of emission of electromagnetic radiation from the targets. The Daventry experiment was an effort in this direction that gave birth to Radio Detection and Ranging (RADAR) radar concept.

For the detection of non-radioactive targets, the target is illuminated using RF energy emitted by RF transmitter and reflected radio energy is detected using RF antenna and when transmitter and receiver are co-located, configuration becomes mono-static while if transmitter and receiver are located to different locations the configuration becomes bi-static. This necessitates requirement of RF transmitter, RF receiver and target and these components can be located on surface, air or in space leads to large spectrum of radar configurations. Originally radars used mono-static

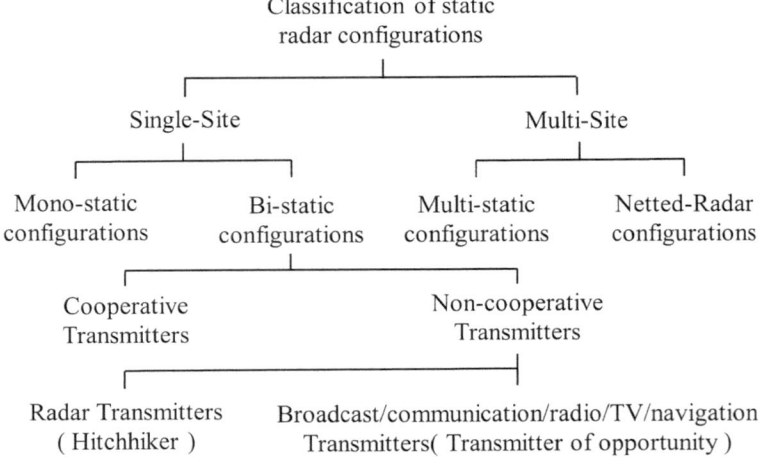

Fig. 9.9 Classification of static radar systems

configuration while separation of transmitter and receiver has great advantage that receiver is passive and can hide its identity and becomes suitable for covert operations and such bi-static configurations provide better immunity to the purposeful change of target shapes designed for hiding.

9.7.1 Classifications of Various Radar Configurations

The variety of configurations are possible about whether each element of radar is stationary of moving, whether transmitter uses continuous wave or pulsed radiation, whether transmitter is cooperative or non-cooperative, whether transmitter and receivers are co-located or distinctly located or whether it uses active or passive configurations.

In case of static configurations, the transmitters and receivers are stationary and it is important whether transmitter and receiver collocated are distinctly located. If both transmitter and receiver are collocated then configuration is called mono-static while if both transmitter and receiver are located are different locations then configuration is called bi-static configuration. The multi-static configuration requires multiple widely spaced transmitters and receivers working together. The static configurations are classified as shown in Fig. 9.9.

9.7.1.1 Active and Passive Radar Configurations

The active radar [254] set consists of two types of resources called transmitting device or transmitter and receiving antenna device or receiver, necessarily paired together to perform the radio detection function to locate the targets under consideration, and when such set of configuration employs the dedicated transmitter, then it is called has *active radar* and hence there can be active configurations in both mono-static and bi-static cases of the radar.

However, if radar configuration does not have its own dedicated transmitter but makes use of other external non-cooperative transmitting sources such as transmitters of other radars, radiating targets such as ships and aeroplanes and public broadcast transmitters then such radar configurations are called *passive radars* [254]. The methods of illumination of targets by such *non-cooperative transmitters* are usually called *transmitters of opportunity*.

In case of active radar, pair of both transmitter and receiver belongs to the same authority that are synchronised and works in co-operative manner and such transmitters which are treated as co-operative transmitters may be collocated or spatially distributed. While in case of passive radars, non-cooperative transmitters such as totally unknown transmitters of other radars or public broadcast transmitters are used. Mono-static or bi-static radar with its own friendly synchronised transmitter is the example of active radars, while radars which do not have their own source of radio transmitters and totally rely on the use of radiation present in the space are called passive radars.

9.7.1.2 CW and Pulsed Radar Configurations

The radar transmitters may emit continuous waves (CW) or pulsed radiation and accordingly these types of radars are classified as CW radars or Pulsed radars.

9.7.1.3 Stationary and Non-stationary Radar Configurations

The complete radar system can be stationary or moving when installed on moving platforms and such radar configurations are classified in Fig. 9.10.

9.7.1.4 Mono-static Radar Configurations

In mono-static systems also called active radars where transmitter (Tx) and receiver (Rx) are collocated or closer to each other as shown in Fig. 9.1b. The mono-static radar can be implemented using single bi-directional antenna serving the purpose of both transmission and reception of the signals and requires duplexer to switch the antenna between transmitter and receiver. The mono-static radar can also be implemented using separate collocated transmitter and receiver with their own antennas. In mono-

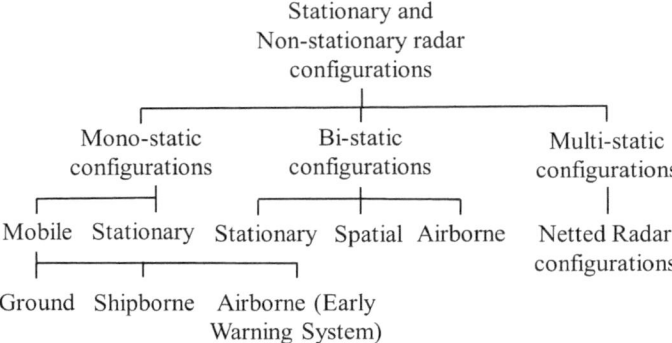

Fig. 9.10 Classifications of radar systems installed on stationary and moving platforms

static case, transmitter is synchronised with receiver and measurement of received signal delay gives an idea of target range.

9.7.1.5 Bi-static Radar Configurations

The bi-static (BS) configuration [255, 256] uses a pair of spatially separated transmitter and receiver for detection of moving targets such that all of these three elements form triangular geometry in which distances between transmitter and target(d_{tr-t}), target to receiver (d_{t-r}) and transmitter to receiver (d_{r-tr}) from the sides of the triangle.

Although, bi-static configuration uses two separate entities called transmitter and receiver at different locations or spatially separated, always represent a single entity and therefore can not be called multi-site or multi-static radar systems. However, bi-static radar do have basic element of multi-static radar system in reality.

Early bi-static radar system used single frequency continuous wave (CW) transmitter which was spatially separated from the receiver, was dependent on principle of Doppler frequency shift of continuous wave transmission (CW) when reflected from the moving target and received by the receiver.

While considering the geometry of bi-static radar, the straight line connecting transmitter and receiver or straight distance between transmitter and receiver is called bi-static baseline (L) while angle between the vectors between transmitter to target and target to receiver is called bi-static angle (β) and depending on this bistatic angle β, the various regions defined are called as pseudo-mono-static region when β lies between 0° and 50°, bistatic region when β lies between 50° and 150° and forward scatter region when β lies between 150° and 180°. The definitions of these regions play an important role in radar-based activity monitoring (RAM).

When both are parts of the radar that is transmitter and receiver which are of same agency then transmitter is co-operative while if transmitter of another agency may be

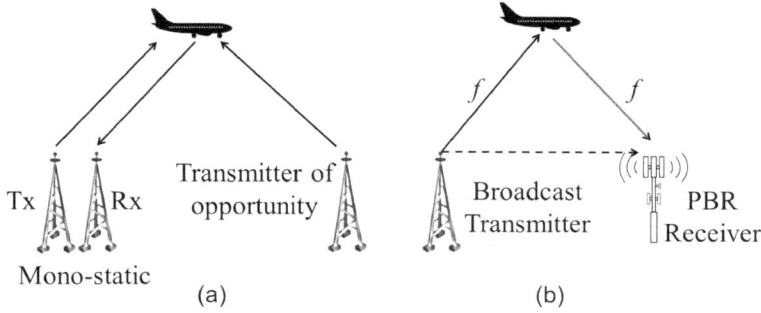

Fig. 9.11 **a** Bi-static Hitchhiker radar system and **b** Passive bi-static radar system

hostile is called non-cooperative or hostile transmitter and both of such transmitters can be used for detection of target are called as transmitters of opportunity.

The bi-static radar can use its own dedicated co-operative transmitters or non-cooperative transmitters of other radar and public broadcasting transmitters like FM broadcast stations, television, cellular systems, GPS and naval transmitter systems as the transmitters of opportunity [257].

The passive bi-static radar (PBR) configuration [258] makes opportunistic use of non-cooperative transmitter of another radar or non-cooperative public broadcast transmitters but when passive bi-static configuration makes opportunistic use of non-cooperative transmitter of another radar it is called as bi-static hitchhiker or parasitic radar [259]. The configurations of both bi-Static Hitchhiker and passive bi-static radar systems are shown in Fig. 9.11. Before the World War II, the Bi-static radar experiments using continuums wave (CW) for aircraft detection were simultaneously conducted in many European countries like Britain, Germany, France and Italy while countries active outside Europe were Russia, America and Japan.

9.8 Complex Radar Configurations

A radar for passive target detection is always configured using a pair of transmitter and receiver and various configurations such as transmitter and receiver at one place (mono-static case) or can be configured or a pair of transmitter and receiver located at distinct places (bi-static case) or one transmitter and multiple receivers (multi-static case) or multiple transmitter and single receiver (a multi-static case) or multiple transmitters and multiple transmitters (fully multi-static case) and each configuration can act as a particular radar site. The use of any combination such configurations working together in a coordinated manner is called multi-site radar system. In order to work out further configurations for proper presentation, let us designate such configurations using some symbol as shown in Fig. 9.12.

Fig. 9.12 Radar
configuration symbols

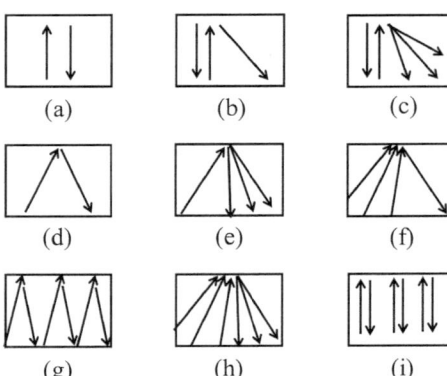

(a)   (b)   (c)

(d)   (e)   (f)

(g)   (h)   (i)

Fig. 9.12a represent ***mono-static*** configuration with its own dedicated transmitter and hence it has be necessarily active radar configuration. Fig. 9.12b is a combination of both mono-static and one bi-static configurations together in which receiver other than receiver of mono-static configuration uses transmitter of mono-static radar as a transmitter of opportunity and hence bi-static configuration is passive bi-static configuration while mono-static is active configuration. This configuration is also called ***bi-static Hitchhiker*** configuration. The Fig. 9.12c consists of multiple passive bi-static configurations along with one active mono-static configuration also called ***multi-static Hitchhiker*** configuration or ***multiple bi-static network*** case. The both configurations of Fig. 9.12b, c are Hitchhiker radar configurations. Fig. 9.12d represents ***active bi-static*** or ***passive bi-static*** configuration depending on whether dedicated transmitter or transmitter of opportunity is used. Fig. 9.12e, f are ***multi-static*** configurations. The implementation in Fig. 9.12e consists of single transmitter of opportunity and connected with multiple receivers in bi-static configuration while Fig. 9.12f implements multi-static configuration using multiple transmitters connected with single receiver in bi-static modes. Fig. 9.12g represents multiple independent multiple connected bi-static configurations representing ***multiple bi-static network*** case while Fig. 9.12h fully ***multi-static network*** configuration using multiple transmitters of opportunity connected to multiple receivers in MIMO configuration. Fig. 9.12i also represents the ***multiple mono-static network*** or ***netted mono-static radar network*** case.

9.9 Multi-site Radar Configuration/System

Multi-site radar system [260] covers universal larger domain of radars working in cooperation and covers both multi-static and netted radar systems which is defined as

Fig. 9.13 Multi-site radar
system

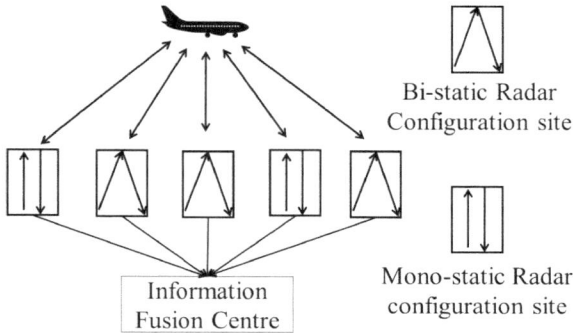

Bi-static Radar
Configuration site

Mono-static Radar
configuration site

Information
Fusion Centre

Definition 9.9.1 Definition of Multi-site Radar System (MSRS): A Multi-site radar
system (MSRS) is a radar system that includes several spatially separated transmitters, receivers and (or) transmitter-receiver stations where target information from
all transmitters and receivers is jointly processed and fused together and jointly processed at Information Fusion Centre (IFC) that is connected to all transmitters and
receivers using Data Communication Lines (DTLs). The multi-site radar system is
shown in Fig. 9.13.

9.9.1 Multi-static Radar System

Perhaps there is surprising perception that there is no well accepted definition of
multi-static radar and subject had been revisited by many authors in past and various
terms used in the definition are rather confusing when one deals with bi-static radar
configuration where spatial separation between transmitter and receiver is significant. The bi-static radar is active if it is not completely reliant on deliberate use of
electromagnetic emission from the target itself but can use its own dedicated spatially
separated transmitter-receiver pair.

Multi-static radar is variation of bi-static radar and uses multiple antennas at
separate locators in the form of either one transmitting antenna pairing with multiple
receiving antennas or one receiving antenna pairing multiple transmitting antennas all
at separate locations in bi-static modes. The transmitters used in such configurations
can be the transmitters of other radar system working in either mono-static or bi-
static modes or public broadcast transmitters or its own dedicated transmitter or
transmitters. The use of transmitting sources other than its own transmitting sources
is called passive radars or parasitic radars or piggyback radars or passive coherent
location radars. In the most generic way, multi-static radar system is defined as
follows.

Definition 9.9.2 Definition of Multi-static Radar (MS) [261]: Multi-static radar systems are the systems of a generally high complexity where radar transmitter and

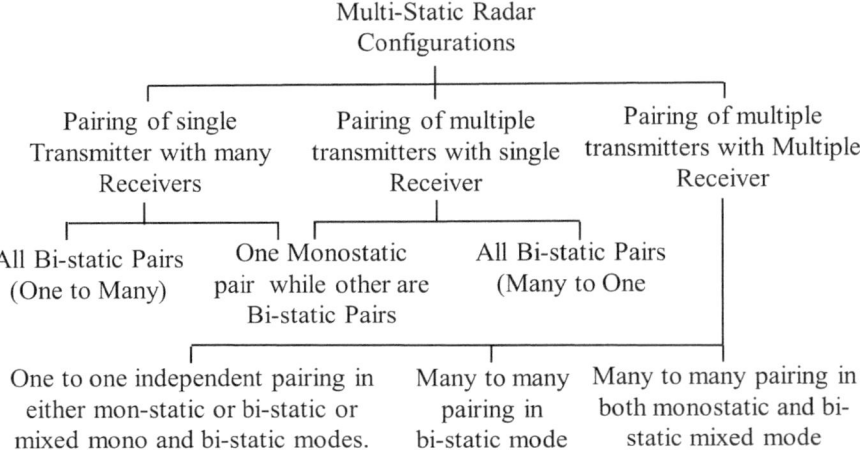

Fig. 9.14 Classifying transmitter and receiver pairing in multi-static radars

receiver subsystems are employed for common spatial coverage in a coordinated manner having two or more receiver sites in which data from each site is combined at a central location

9.9.1.1 Classification of Multi-static Radars

The multi-static radar configurations [262] cater to one to many and many to many pairing of radio transmitter and receiver pairs in which one transmitter pairs with multiple receivers or many transmitters pair with single receiver or many transmitters pair with many receivers and vice versa. In many to many configurations, it may be the simple one to one interactions in the form of independent pairs together or cross-coupled interactions and accordingly the classification chart of pairing of transmitters and receivers for multi-static radars is shown in Fig. 9.14.

9.9.1.2 Multi-static Radars: One to Many Transmitter-Receiver Pairing

In case of one to many interactions, either transmitter or receiver pair with their respective multiple counterparts where four topological configurations can be worked out as shown in Fig. 9.15. The configurations in Fig. 9.15a, c are bi-static configurations either active or passive depending on whether transmitters are either dedicated transmitters or transmitters of opportunity. In one transmitter to many receivers pairing configuration, all receivers are always made to accept the signals related single frequency of the transmitter while in case of many transmitters to -one receiver pairing configuration, single receiver is made to receive signals related to transmitters of

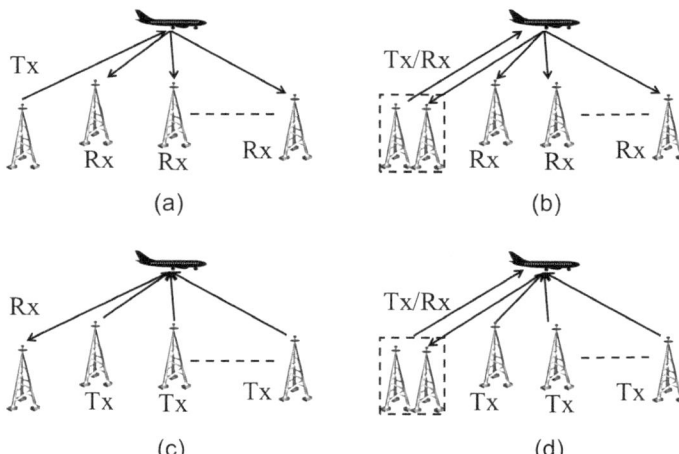

Fig. 9.15 One transmitter-many receivers and many transmitters-single receiver multi-static configurations

different frequencies and hence receiver has to multichannel. Fig. 9.15a represents bi-static configurations involving single transmitter pairing with multiple receiver and hence such multi-static configuration operates on single frequency transmitted by the transmitter under consideration and all receivers are single channel while the configuration in Fig. 9.15c, since single receiver is required to accept the signals related to the multiple frequencies all of its counterpart transmitters, receiver has to be multi-channel. Fig. 9.15b, d involves combination of both mono-static as well as bi-static radar configurations. The configuration Fig. 9.15b is typical Hitchhiker configuration while the configuration in Fig. 9.15d forms bi-static configurations between receiver of mono-static radar and transmitters other than the transmitter of mono-static radar.

9.9.1.3 Fully Multi-static Radars: Many to Many Transmitter-Receiver Pairing

Fully multi-static radar consists of many transmitters and many receivers in which each transmitter or each receiver can pair with to some or all of its counterparts as shown in Fig. 9.16. Fig. 9.16a is fully multi-static radar implementation using all bi-static pairs and each receiver has to be multichannel and Fig. 9.16b is fully multi-static radar implemented using both mono-static and bi-static configurations where all receivers are multichannel except receivers of mono-static configuration which cater to single channel.

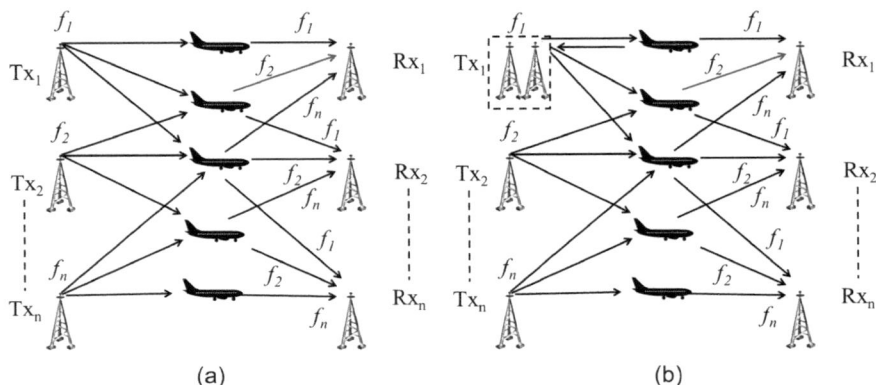

Fig. 9.16 a Fully multi-static radar using only bi-static configuration. **b** Fully multi-static configuration using both mono-static and bi-static configurations

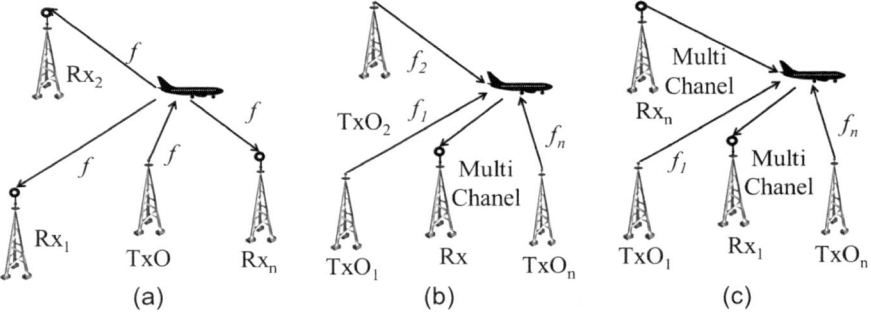

Fig. 9.17 Passive multi-Static radar network configurations

9.9.1.4 Multi-static Passive Radar Topologies

The various configurations of passive multi-static radar networks [263] can be worked out as (a) one transmitter of opportunity (TxO) and many receivers, (b) one receiver and many transmitters of opportunity (TxOs) and (c) multiple transmitters of opportunity(TxOs) and multiple multi-channel receivers. These configurations are shown in Fig. 9.17.

9.9.2 Networks of Radars

The networks were used to increase the accuracy and coverage of the operations and various types of networks such as multi-static (M-STAT), Multilateration (MLAT)

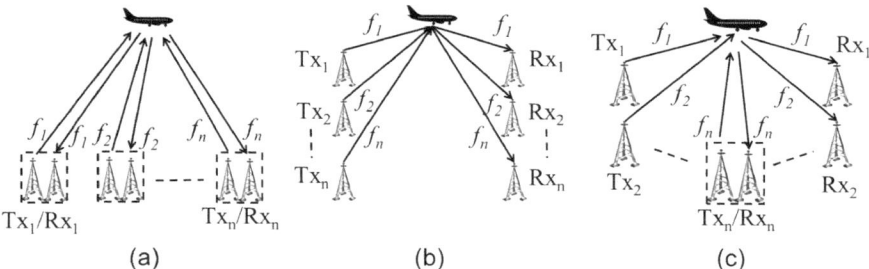

Fig. 9.18 Independent Mono-static, bi-static and mixed mono-bi-static radar networks or netted radars

and network of radars were being adopted. It consists of several radars linked together to improve the accuracy and coverage due to union of individual coverage together.

9.9.2.1 Netted Radars: Network of Mono-static or Bi-static or Mono-static and Bi-static Radars

It is actually network of radars using mono-static or bi-static or both mono-static and bi-static configurations as shown in Fig. 9.18. In this case also we can make the distinction of given transmitter and receiver pair as active or passive depending on whether dedicated transmitter or transmitter of opportunity are used.

9.9.2.2 Multi-lateration (MLAT) System

The multilateration (MLAT) system is based on the principle of estimation of position and time delay of the moving target. The time difference of moving target is estimated using Time Difference of Arrival (TDOA) implemented by various receivers and central unit connected using links. TDOA is widely used in the location of moving objects like aircrafts. Multilateration steps consist of use of Mode A, C and S transponders for mode ACS interrogation, Mode A/C/S Reply for identification of friend of foe (IFF), TDOA processing, hyperbolic positioning and finally aircraft portion is displaced. The concept of multilateration is shown in Fig. 9.19.

9.10 First Generation Radar Development

The countries like Germany, United Kingdom and United States played important role in radar developments that ended up with using similar type of basic conceptual radars before _1940_, shown in Fig. 9.20 called first-generation radars. The first generation radar relies on a single parameter called position while range is calculated

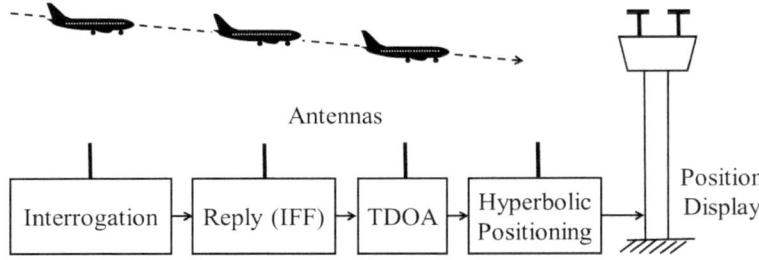

Fig. 9.19 The concept of multilateration

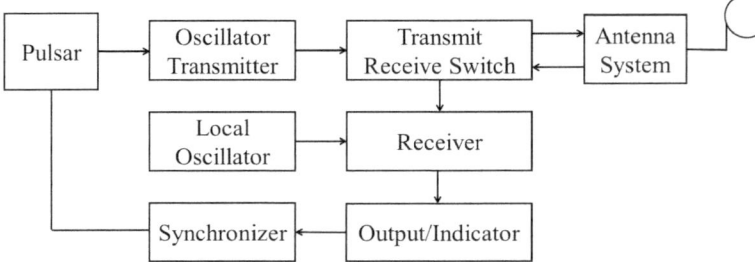

Fig. 9.20 Basic radar system prior to World War II

by measuring time gap of transmitted pulse and received echo pulse or radio wave that is reflected by the target object and direction of the object is obtained from the angular direction of an antenna. The invention of radar is a clear case simultaneous developments in many countries, however, in the pretext of secrecy, it is difficult to ascribe an individual who invented the radar. The most influential publication entitled "Detection and Location of Aircraft by Radio Methods" came from British scientist Robert Watson in *July 1935* who was consistently able to locate an aircraft at the distance of 90 miles. His system of radars was called "Chain Home" operating on the frequencies of 22–50 MHz and was in operation for 24 h in *September 1938*.

9.10.1 Watson-Watt's Chain Home (CH) Radars

In *May 1935* Watson-Watt, Wilkins and a small team of scientists moved on to a long narrow projected portion of land in to ocean or projected ridge at Orford called Orford Ness [264] on the Suffolk coast of Britain to conduct a series of historic experiments over the sea that lead to the world's first working 'RADAR' system. The Orfordenss separated from mainland Aldeburgh by River Alde was connected only using small ferry from Orford Quay. Earlier in *1913*, during World War I, Ministry of Defense acquired major portion of Orfordness for the construction of airfield and in *1933*, a Bomb Ballistics Building was constructed at this site considered to ideal site for a team of Watson-Watt when they moved on this site in *1935*. The Ionosphere Research Station to cover the radio detection facility to be used for radio direction finding (RDF) was established and first ever 250 ft (76.2 m) radar antenna masts were built in 1935 by Harland and Wolf industry of Belfast BT3 9DU, Northern Ireland (UK).

It soon became apparent that Orfordness was inadequate for further research and hence Bawdsey Manor Estate near Deden river was purchased for £24,000. In *February 1936*, the research scientists occupied Bawdsey Manor House and the stables and outbuildings were converted into workshops. 240 ft wooden receiver towers and 360 ft steel transmitter towers were built and Bawdsey became the first Chain Home Radar Station. As a result of his success, subsequently in *August 1936*, Watson Watt was made a superintendent of Bawdsey Research Station [265] near Felixstowe in Suffolk coast belonged to Air Ministry that paved a for creation of chain of radars along the east coast of England. The Bawdsey Research Station near Suffolk coast is shown in Fig. 9.21a while Bawdsey radar station is shown in Fig. 9.21b

This system was also called Chain Home radar system was typically a ground controlled intercept Radar (GCI) that became most vital part of UK's defence against Germans in "Battle of Britain" that served as an early warning system for Fighter Command of Britain. In the same year (*1936*), three radar stations were setup at Bawdsey Manor, Canewdon and Dover while additional two stations were set up Gt Bromley and Dunkirk in *1937*. These five stations were set up to protect Thames Estuary also called Thames Estuary Chain to test the radar principle. The geographical locations of these Chain Home radars are shown in Fig. 9.21d. The RAF (Royal Air Force) Bawdsey Radar station was the first fully operational Radar station, made operation on 24 September 1937. Fig. 9.21c shows the Chain Home transmit tower at Great Baddow Last surviving Chain Home transmit tower that was originally sited at Canewdon moved from Canewdon to Great Baddow.

As described above, the initial setting up of Chain Home started taking place on south eastern coast of England and spread over entire coast line of England. The most prominent region was Thames Estuary where chain home radars were established at Canewdon [266], Dover, Dunkirk and Gr Bromley. The Chain Home Radar Stations at Canewdon [266] and Dover were one of the firsts in the history built in 1936 and within the span of six months and just before the start of World War II, 21 Chain Home Radar stations were built and made operational. The chain home radar towers at Swingate, Dover, are shown in Fig. 9.22. The east coast chain home (CH) radar system used four transmitters and four receivers with intension to use duplicate pair of Chain Home transmitters type T.3026 and of pair of receivers at a time with intension to keep other pairs as standby pairs in case of failure. Both main and standby pairs used four frequencies from 20 to 30 MHz frequency band. These radars used pulses of 20 μs with pulse repetition frequency of 25 and 122.5 pulses per second at peak pulse power of 350 kW. The four frequency plan was also abandoned later.

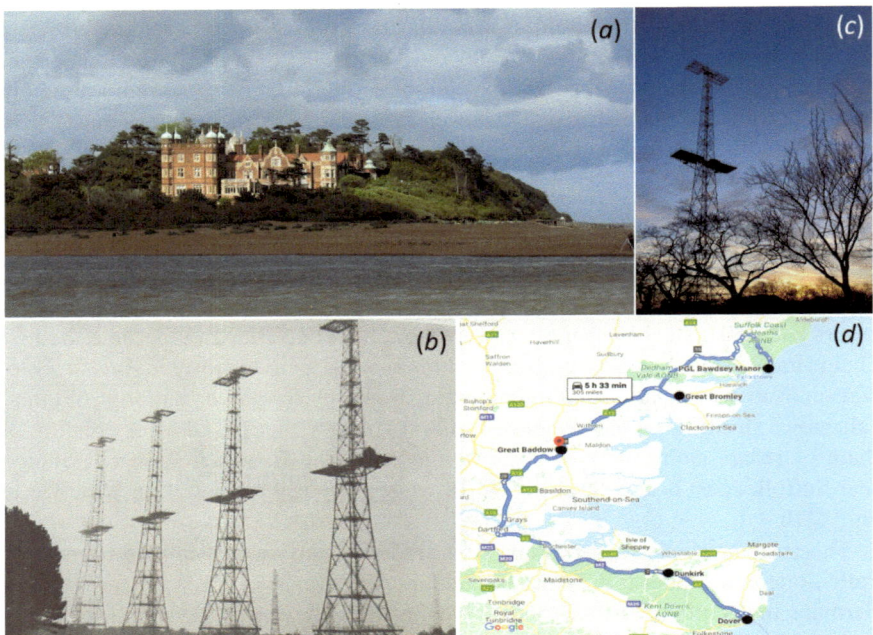

Fig. 9.21 **a** Bawdsey Manor R&D Centre on Suffolk coast an image produced by the permission of Bawdsey Radar Trust, **b** Bawdsey Chain Home Radar station. Courtesy: Gordon Kinsey, 'Bawdsey, Birth of the Beam', Terence Dalton Limited, Lavenham, Suffolk 1983 and Colin Statham & Anne Stobbs, Radar - A Wartime Miracle, Sutton Publishing Limited, 1996, **c** Chain Home transmit tower at Great Baddow Last surviving Chain Home transmit tower that was originally sited at Canewdon moved from Canewdon to Great Baddow an image produced by the permission of Bawdsey Radar Trust and **d** Prominent Thames Estuary Chain Home sites Courtesy: Google Maps

9.10.1.1   Constructional Details of Transmitting and Receiving Antennas of Chain Home Radars

The chain home radars installed on the east coast consisted of four steel transmitting towers(which were later reduced to three) and receiving arrangement consisted of four wooden towers in rhombic formation located about hundreds of yards away from the transmitter buildings [267]. The sky over the area to be kept under surveillance was virtually floodlit using RF pulsed energy with the help of transmitting towers while reflected echo pulses or scattered pulses from the target aircraft received using set of crossed dipole antennas connected to low noise and high gain receiver to displayed on CRT screen. The construction details of transmission antennas are shown in Fig. 9.23.

   The receiving antennas mounted on each wooden tower consists of three sets of antennas set1 and set 2 are similar and consist of two sets of centre-fed horizontal crossed dipoles which are aligned to E-W and N-S direction. E-W dipoles are called

(a)

(b)   (c)

Fig. 9.22 Photographs of the status of the Chain Home Radar site at Swingate, Dover, Kent, UK over the passage of time **a** The Radar towers of Dover C-H station seen from the coast of France in 1940, Credit: Late Art Cockerill, http://richardgilbert.com **b** Remaining three towers in 2010 Credit: Bob Cromwell, https://cromwell-intl.com, a website maintained by Bob Cromwell **c** Photograph of remaining two transmitting towers in 2017, Credit: David Lovell, Historic England

Fig. 9.23 Chain home radar transmitting antennas

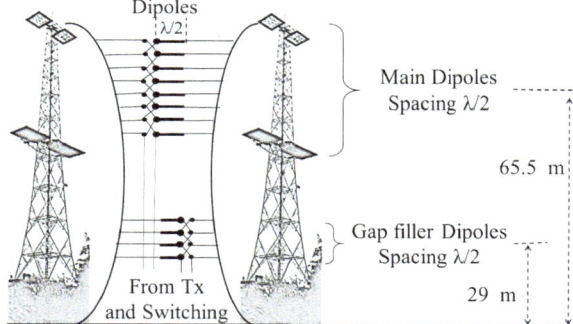

X dipole while E-W dipoles are called are Y dipoles. The switched reflectors are placed on north side of E-W dipoles and west side of N-S dipoles while set 3 consists of stack of two single dipoles with non-switched reflectors. The construction details of receiving antennas are shown in Fig. 9.24.

Fig. 9.24a shows rhombic formation of wooden receiving towers, and Fig. 9.24b shows the arrangements of dipole antennas and reflectors on the wooden receiving towers.

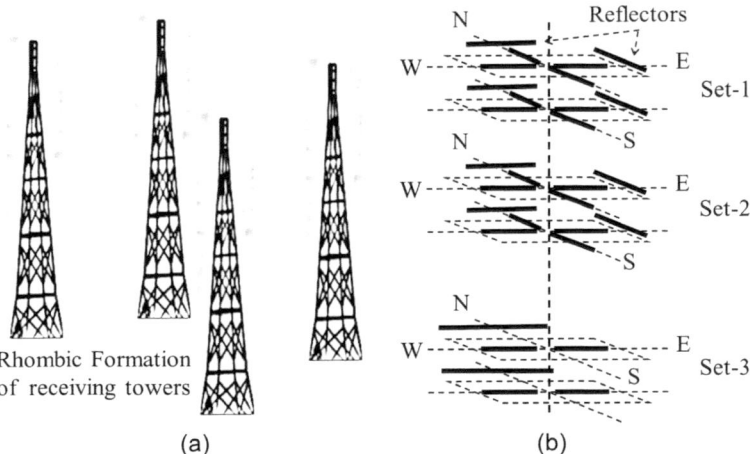

Rhombic Formation
of receiving towers

(a)   (b)

Fig. 9.24 Chain home radar receiving antennas

9.10.1.2 Thames Estuary Chain Home Radars Become Operational on Full Time Basis

In *1938*, the five Themes Estuary chain home radars started working on full time basis, however, due to lack of networking, these systems exhibited some difficulties in communicating useful information to its pilots and hence radars in Britain and Scotland covering entire airspace were networked to become world's first wide area network of the radars known as Dowding System.

9.10.1.3 Hugh Dowding Hierarchical System

Air Chief Marshal Lord Hugh Dowding, commander-in-Chief of Fighter Command, worked tirelessly towards the development of comprehensive air defence system in the form of networks of radio direction finders (RDFs) (which were later called radars) and all the chain home radars were connected using dedicated telephone lines to form the single image of entire UK airspace for directing the defence aircrafts using hierarchical reporting controlled by Fighter Command Headquarters (FCHQ) central filter room at Bentley Priory This hierarchical system was called Dowding system is shown in Fig. 9.25. Where filter room (F) is connected to group rooms (G), group rooms are connected to sector rooms (S) and sector rooms are connected to Ground Control Interception (GCI) Radars. In addition, filter room was also connected to various posts of Royal Observer Corps for visual tracking and also to individual chain home radars of the networks formed using chain home radars. The complete system was at his disposal after 1939 for use. The system played a crucial role in

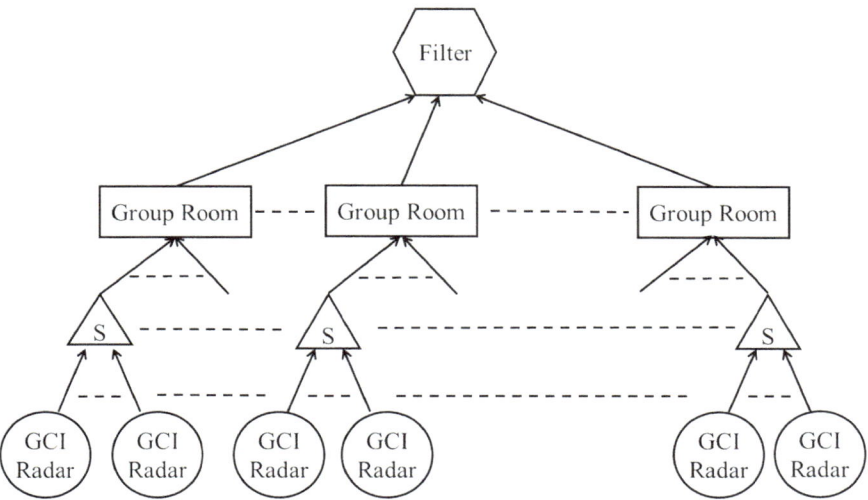

Fig. 9.25 Hugh Dowding hierarchical system

Fig. 9.26 Frame coils for
direction finders, Courtesy:
HMS Collingwood Heritage
Collection

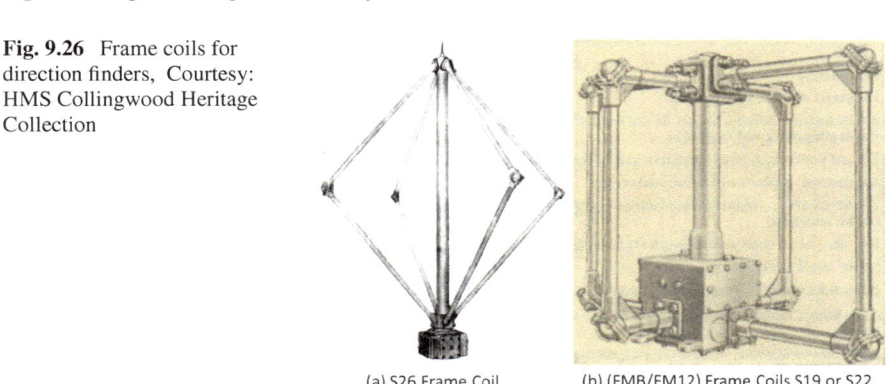

(a) S26 Frame Coil   (b) (FMB/FM12) Frame Coils S19 or S22

Battle of Britain from *July to October 1940* and had a great success against German
Luftwaffe.

9.11 Progress in Frame Coils for High Frequency Direction Finders

During *1937*, the direction finder outfits called FA, FC and FH used frame coils fitted
on the tip of mast head. The frame coils S16 and 17 were typically used with FA
(S16 or S17),FC(S16 or S17) and FH(S16) outfits (Fig. 9.26).

9.12   Field Trials of HF/DF Equipments and Loop Antennas on Navy Vessels

The further trials were conducted on HMS Argus vessel having no masts or rigging in 1938 for the purpose of establishing the effects of other factors than re-radiation. By this time improved designs of HF/DF (Goniometer) were available and just before the outbreak of World War II, installations of this equipment were done on many warships like HM Battlecruiser Renown with fixed loops, HM Cruiser Aurora with two rigid loops and HM Cruiser Manchester with a rotating Frame Coil and soon after World War II broke, Cruisers Euryalus and Bonaventure were fitted with rotating frame coil antennas.

The most important discoveries of *1939* was the use of antennas operating at highest point, operating at frequencies above 2MHz (Range of frequencies from 2MHz to 30MHz) to minimize errors due to surrounding which played a crucial role in development of most operationally viable FH1 system by HM Signal School in *March 1941* that was installed on HM Destroyer HESPERUS.

9.13   Starting of Developments of Airborne Radar Systems

Early radars were operated at relatively low frequencies and thus were relatively immune to most types of natural clutter. The clutter refers to unwanted radio frequency (RF) echoes returned from targets which do not have positive contribution in detection of the target and interfere with the useful signals. The main challenge was to discriminate useful weaker target echo signals when clutter was far more stronger than the useful echo signals received from targets of interest and coherent signal processing plays an important role in such cases of clutter. The IEEE Standard (Std 686-*1990*) defines coherent signal processing [268] as echo integration, filtering or detection using the amplitude of the received signals and its phase referred to that of a reference oscillator or to the transmitted signal.

The earlier gigantic "Chain Home" radars masts served purpose of early warning to Britain's fighter defence, however, the airborne radar developments required smaller antennas and dipole antennas offered promising solution that operates best at half wavelength of radar beam. The availability of short wave sources which can provide sufficient power was lacking and need was to develop high power shortwave vacuum tubes.

9.13.1   Background of Development of 1.5 m Radar Systems

The chain home radars was an effective weapon during day time and in *1936*, Henry Tizard wrote to Dowding commenting on accuracy of Watson Watt's radar as sufficient to find the position of enemy aircraft during day when pilots can visually locate the enemy target but insufficient during night time and that is why German were forced to attack during night time. This prompted the need of inclusion of night operation in aircraft-based radar system and owing to this Watson-Watt immediately responded with splitting of operations in to two groups one group can look after existing chain home radars and other group can focus on night interception aspects.

9.13.2   Splitting of Responsibilities Between Wilkins and Bowen

The responsibility of chain home radars was entrusted to Wilkins and work on night interception radar was assigned to Dr Edward George 'Taffy' Bowen. In *August 1936*, Watson-Watt, who was working as Superintendent of Air Ministry Research Establishment (AMRE) at Bawdsey, formed a group under the leadership of Dr Edward George 'Taffy' Bowen to manage the development of air intercept (AI) radar. In consultation with Royal Air Force, Dr Edward George 'Taffy' Bowen pen downed the limiting specifications of airborne radar in terms of mass (90 kg max), volume (0.22 cubic meter max.) and power (500 W max.) as the basic requirements.

In *October 1936*, on the basis of Watson Watts work at Bawdsey, British army got interested in the work of radio detection and team led by E. T. Paris and Albert Beaumont Wood set up Military Applications Section at Bawdsey also called "Army Cell". In the same year in *1936*, the term called Airborne Interception radar (AI) radar came into existence when Bawdsey Manor research center considered an option to fit radar into an aircraft which led to the work of Airborne.

9.13.3   Bowen's Work Towards Airborne Radar

It was then search for the suitable radio receiver, Edward George 'Taffy' Bowen aware of receiver that was being used for receiving BBC TV transmission produced by a company called EMI, thought of using this receiver for which he built a transmitter operating on 6.7m wavelength (45 MHz in VHF Range) to be used on the ground, while receiver was to be mounted on the aircraft. In Autumn of *1936*, the receiver was mounted on Heyford bomber to make it bi-static configuration with ground transmitter and resultant system was called Radio Direction Finder (RDF-1R) that detected aircraft target up to the range of 19 kms.

Fig. 9.27 Western Electric's
316A/VT-191 UHF
transmitting triode,
'doorknob' style vacuum
tube valve(Western Electric,
RMA record 339) *Source*
ase-museoedelpro.org

9.13.3.1   Efforts for Antenna Size Reduction for Complete Airborne Applications

The further reduction of antenna size was highly desired by use of further higher frequencies. The Western Electric worked towards the development of high power valve to fill up this gap and came out with Electric Giant Acorn 316A vacuum tube valve [269, 270] that generated 100W at 1.25 m wavelength ($\backsimeq$ 240MHz in VHF range) in *1937* is shown in Fig. 9.27.

This valve was used to generate a pulse of 2–3 μs repeated at the rate of 1 ms with a peak power few hundred watts. The invention allowed the complete radar system to be airborne unlike earlier case where transmitter was located on the ground while receiver was airborne. Using this valve, first British airborne radar was developed and on *March 1937*, the system was installed on Heyford bomber that detected vessels and infrastructure facilities of Harwich.

9.13.3.2   Air to Air Mode of Operation

During *August 17, 1937*, for the purpose of maritime surveillance this radar was used as airborne surface vessel (ASV) detection radar by installing it onboard of Avro Anson that demonstrated its ability to detect an aircraft at the distance of one mile in air to air mode and also detected the ships at the distance of 3 miles in air to surface mode. In fact, this experiment demonstrated the fact that it can be used as both airborne Interception radar (AI)/indexAirborne Interception (AI) radar and airborne surface vessel (ASV) radar. In fact, this experiment provided split mode operation at place which otherwise considered to be two separate tasks executed at two distinct and different locations. It also provided merging of distinction between Airborne Interception radar (AI) and airborne surface vessel (ASV) radar.

9.13.3.3   Practical Use of Airborne Radar

On *September 4–5, 1937*, Avro Anson K6260 with Bowen's radar installed, partici-
pated in an aero-naval exercise and received strong echo signals from Cruises, battle
ships, destroyers and aircraft carriers.

In an effort to further to improve the sensitivity of airborne radar, the wavelength
was increased from 1.25 m (240 MHZ) to 1.5 m (200 MHz) in *August 1939* and radar
first hand built AI MK.I came in to existence and six of these were installed on six
Blenheim fighters IVF while ASV MK.I radars were installed on few costal command
aircrafts.

By *January 1940*, at the beginning of World War II, ASV Mk.1 radars operating
on 1.5m wavelength (200 MHz)were installed on nearly 49 aircrafts (14 Hudson
aircrafts and 50 Sunderland aircrafts).

9.14   Airborne Television Systems for Military Applications

On *21 July 1939*, John Logie Baird worked for French air Ministry for ground surveillance from the aircraft and was asked to develop television system that can locate
enemy movement on the ground using television imaging using his equipment called
"Air craft to ground television equipment" [271]. His experiment consists of television equipment installed on airplane that picked up the pictures from aircraft and
transmitted to a vehicle on ground having receiving equipment. A series of moving pictures were captured on 16 mm film which was then rapidly processed and
transmitted to the vehicle standing on the ground.

9.15   Gathering of Information by Germans Using Graf
         Zeppelin Airship (LZ-130)

During *August 1939*, Graf Zeppelin airship (LZ-130) [272] was spying over north
sea could see and detect signals from CH radars but made a gross mistake to conclude
that signals may be due to defective insulators on the National Grid or due to radio
communication or due to radio navigation rather than radar.

9.16   World War Begins

The World War II started in *September 01, 1939* and continued till *September 02, 1945*
and during this period, military defence and communication systems gained much

more importance and by *1939*, Britain, Germany, France, Hungary, Italy, the United States, Japan, Netherlands, Russia and Switzerland all had operational radars.

9.17   Conclusion

This chapter is devoted for the development of wireless systems just before the World War II, in which major contributions are focused on how to detect moving of stationary enemy targets and defence from enemy attacks. The major contributions were from Watson Watt and his groups working under him.

References

1. T.P. Shaffner, *The Telegraph Manual: A Complete History and Description of the Semaphoric, Electric and Magnetic Telegraphs of Europe, Asia, Africa, and America, Ancient and Modern* (D. Van Nostrand, No. 192, Broadway, New York, 1867)
2. G.B. Prescott, *Electric Telegraph*. Ticknor and Fields, Boston, Elecrographed and Printed by Welch, Bietlow & Co., Unversity Press, Cambridge, 4th edn., rev. and enl. edition, 1866. Reprinted September, 1972 by: Frank Jones Hwy. #25, Box 73 Cottontown, Tenn. 37048
3. R.W. Burns, *Communications: An International History of the Formative Years* (Institution of Electrical Engineers, Michael Faradays House, Six Hills Way, Stevenage, Herts, SG1 2AY, London, UK, 2004)
4. Lokilech, Optischer telegraf, von claude chappe auf dem "litermont" bei nalbach in deutschland. https://en.wikipedia.org/wiki/Telegraphy/media/File:OptischerTelegraf.jpg, 14 May (2006)
5. M. Fowler, Historical beginnings of theories of electricity and magnetism. Technical report, Department of Physics, University of Virginia (1997)
6. B. Franklin, Experiments and observations on electricity at Philadelphia in America by Benjamin Franklin. Printed for David Henry and sold by Francis Newsbery, at corner of St Paul's Church Yard, London (1748)
7. C. DuFay, A discourse concerning electricity. Philos. Trans. R. Soc. **38** (December 1733)
8. R. Routledge, *A Popular History of Science*, 5th edn. (George Routledge and Sons, The Broadway, Ludgate, London (UK), 1881)
9. J. Munro, *Heroes of the Telegraph* (The Religious Tract Society, 56, Peternoster Row: 65, St. Paul's Church Yard and 164 Piccadilly, London (UK), 1891)
10. C. Morrison, Electrostatic telegraph. Scots Mag. **16**, 73 (1753)
11. J.J. Fahie, *A History of Electric Telegraphy, to the Year 1837* (E. and F.N. Spon, 16 Charing Cross, London (UK), 1884)
12. L.L. Figuier, *Les Merveilles De La Science Electricite ou Description Populaire des Inventions Modernes* (Librairie Furne, Jouvet et o Editeurs, 5, Rue Palatine, 5, Paris, 1867). https://gallica.bnf.fr/ark:/12148/bpt6k24674j/f11.image
13. L. Magrini, Notizie, Biografiche e Scientificlie su Alessandro Volta. *Atti del Reale Istituto Lombardo*, II, April 15 (1777)
14. Annonymous Author, Notizie, Biografiche e Scientificlie su Alessandro Volta. Journal de Paris (150), May 30 (1782)
15. R. Routledge, *Lomond's Invention*, vol. I, 4th edn. (Arthur Young's Travels, 1787)
16. Bockmann, Versuchtiber telegraphic und telegraphen. Technical report, Karlsruhe (1794)

17. J.D. Reid, *The Telegraph in America, Morse Memoial* (Wsex Polhenus Publisher, 102 Nassau Street, New York (USA), 1879)
18. T. Cavallo, *A Complete Treatise on Electricity, in Theory and Practice with Original Experiments*, vol. III, 4th edn. (1975)
19. Reizen's telegraph. Voigt's Mag. **IX**, 1 (1794)
20. Reizen's telegraph. Salva's Telegraph **XI**, 4 (1798)
21. J. Priestley, *The History and Present State of Electricity, with Original Experiments*. Printed for J. Dodsley in Pall-Mall, J. Johnson and B. Devenport in Pater-Nofter Row and T. Cadell in the Strand (1767)
22. A. Volta, *Letters of Signor Don Alessandro Volta-on the Flammable Native Air of the Marshes Milan* (Giuseppe Marelli, Italy, 1777)
23. X. Wei, Z. Wang, H. Dai, A critical review of wireless power transfer via strongly coupled magnetic resonances. Energies **7**(7), 4316–4341 (July 2014)
24. W.C. Brown, History of power transmission by radio waves. IEEE Trans. Microw. Theory Tech. **MTT-32**(9) (September 1994)
25. R.W. Simons, Guglielmo Marconi and early systems of wireless communication. GEC Rev. **11**(1) (1996)
26. R. Techo, *Data Communications: An Introduction to Concepts and Design* (Plenum Publishing Corporation, 227, West 17th Street, New York, 10011, July 1980)
27. S.G. Goodrich, *Curiosities of Human Nature* (Brandbury Soden & Company, 12 School Street, Boston, Massachusetts, (USA), 1843)
28. D.L. Sengupta, T.K. Sarkar, Maxwell Hertz, the Maxwellians, and the early history of electromagnetic waves. IEEE Antennas Propag. Mag. **45**(3) (April 2003)
29. H.C. Oersted, *Der Geist in der Natur* (Lorck Verlag, Leipzig, 1850)
30. H.C. Oersted, *Selected Scientific Works of Hans Christian Oersted* (Princeton University Press, 2014)
31. A.-M. Ampere, *Recueil D'Obervations Electro-Dymiques* (Chez Chrochard Libraire, eloitre sain-Benoit, Pres la Rue Des Mathurins, Paris, 1822)
32. R.A. Pizz, *Michael Faraday, Chemist, Today's Chemist at Work, Chemistry Chronicles* (American Chemical Society, May 2004)
33. M. Faraday, Thoughts on ray vibrations. Philos. Mag. **XXVIII**(N188), S.3 (May 1846)
34. M. Faraday, Experimental researches in electricity. III(N188), S.3 (1855)
35. S.W. Thomson, *Mathematical and Physical Papers (Collection from different scientific Periodical from May 1841 till 1882)*, vol. 1 (Cambridge University Press, Cambridge, Shaftesbury Rd, Cambridge CB2 8BS, UK, 1882)
36. J.C. Maxwell, A dynamical theory of the electromagnetic field, Read on December 08,1864. Proc. R. Soc. **155**(Part I), 459–512 (1865). Printed: Taylor and Francis, London
37. J.C. Maxwell, *A Treatise on Electricity and Magnetism*, vol. 1 & 2 (Clarendon Press, Oxford, UK, 1873)
38. H. Hertz, *Electric Waves: Being Researches on the Propagation of Electric Action with Finite Velocity Through Spac* (Dover Publications Inc., New York, USA, 1962)
39. E.P. Dollard, Theory of wireless power. Technical report, KazumotoIguchi Research Laboratory (1986)
40. Electrical Experimenter Nicola Tesla. The true wireless. Technical report, Radio Department, Nicola Tesla Institute, Brasilia, Brasil (May 1919)
41. Electrical Experimenter Nicola Tesla. Experiments with alternate current of high potential and high frequency. Technical report, A Lecture delivered before The Institution of Electrical Engineers, London, UK (February 1892)
42. F.E. Gardiol, About the beginnings of wireless. Int. J. Microw. Wirel. Technol. 1–8 (2011)
43. V.J. Phillips, *History of Technology Series 2: Early wave Detectors*. Peter Peregrinus Ltd. in association with The Science Museum, Printed by A. Wheaton & Co, London, UK (1980)
44. A.G. Lee, The Varley brothers: Cromwell Fleetwood Varley and Samuel Alfred Varley. J. Inst. Electr. Eng. **71**(432), 958–64 (December 1932)
45. K.T. Compton, Biographical memoir of Elihu Thomson, XXI(Fourth Memoir) (1939)

46. D.E. Hughes, Prof. D. E. Hughes's researches in wireless telegraphy. The Electrician **43**, 40–41 (1899)
47. B.J. Hunt, Practice vs. Theory: The British electrical debate, 1888–1891. ISIS **74**(3), 341–355 (September 1983)
48. V.J. Phillips, The Italian navy coherer affair: a turn of-the-century scandal. Proc. IEE Series A **140**(3), 175–185 (May 1993). Reproduced in IEEE, vol. 86, no. 1, pp. 248–258 (January 1998)
49. O.J. Lodge, Method and system for measurement of road profile, August 16 (1898). https://patents.google.com/patent/US609154. US Patent 609154
50. A. Righi, An instrument for generating radio waves: Righi oscillator or spark-gap of 1895. Marconi's Collection Wireless World: Marconi & the making of radio (1896)
51. A.M. Peterson, The world's oldest radio tower: the story of Eiffel tower radio. *Wavescan, Adventist World Radio (AWR)*, November 25 (2012)
52. J. Miller, A chronology of AM radio broadcasting 1900–1960, June 7 (2017). http://jeff560. tripod.com/chrono1.html
53. F. Braun, Uber die Stromleitung durch Schwefelmetalic. Annalen der Physik and Chemie (in German) **153**(4), 556–563 (1874)
54. P. Russer, Ferdinand Braun: a forgotten hero of electronic engineering, in *CALCON*, Calcutta, India (November 2014)
55. J.C. Bose, On polarization of electric ray. J. Asiatic Soc. Bengal (May 1895)
56. D.T. Emerson, The work of Jagadis Chandra Bose: 100 years of millimeter-wave research. IEEE Trans. Microw. Theory Tech. **45**(12), 2267–2273 (January 1998). https://doi.org/10. 1109/22.643830
57. J.C. Bose, On a self-recovering coherer and the study of the cohering action of different metals. Proc. R. Soc. **LXV**(416), 166–172, April 27 (1899)
58. J.C. Bose, On electromotive wave accompanying mechanical disturbance in metals in contact with electrolyte. Proc. R. Soc. **70**(459–466), 273–294 (January 1902)
59. P.K. Bondyopadhyay, Sir J. C. Bose's diode detector received Marconi's first transatlantic wireless signal of December 1901 (the "italian navy coherer" scandal revisited). Proc. IEEE **86**(1) (January 1998)
60. G. Marconi, Looking back over thirty years of radio. Radio Broadcast Mag. **10**(1), 31 (November 1926). http://www.americanradiohistory.com
61. F. Collins, How to construct an efficient wireless telegraph apparatus at a small cost. Sci. Am. Suppl. pp. 21,821, 849–850, February 15 (1902)
62. G. Marconi, *Wireless Telegraphic Communication: Nobel Lecture on 11 December 1909* (Elsevier Publishing Company, Amsterdam, Netherlands, December 1967)
63. N. McEwen, SOS, CQD and the history of maritime distress calls. Telegraph Office Mag. **II**(Issue 1) (1997)
64. Marconi sends first Atlantic wireless transmission, December 12 (1901)
65. Transmission of transatlantic radio signals. IEEE United Kingdom/Republic of Ireland Section, Poldhu, Cornwall, England, 12 December 2001, December 12 (1901)
66. M. Raboy, *Marconi: The Man Who Networked the World* (Oxford University Press, Oxford, UK, May 26, 2016)
67. G. Pelosi, S. Selleri, B. Valotti, From Poldhu to the Italian station of Coltano: Marconi and the first years of transcontinental wireless. IEEE Antennas Propag. **46**(3) (June 2004)
68. In International Wireless Telegraph Convention (Convention Radiotélégraphique Internationale), Berlin, Washington, D.C., USA, November 3, 1906. Department of the Navy Bureau of Equipment, Washington, Government Printing Office in Washington, D.C., USA (1907)
69. E. Braun, S. MacDonald, *Revolution in Miniature* (Press Syndicate of University of Cambridge, University Press, Cambridge, The Pitt Building, Trumpington Street, Cambridge CB2 1RP, UK, 1978, 1982)
70. G.W. Pickard, Means for receiving intelligence communicated by electric waves, November 20 (1906). https://patents.google.com/patent/US836531. US Patent 836531

71. G. Rode, Guide to the History of Lofoten. Technical Report N-8393, Norwegian Telecom Museum, The Loften Public Museums, Sørvågen, Lofoten Islands. http://www.lofoten-info.no/tele-mus.htm

72. L.C. Godara, Introduction to the heterodyne receiving system, and notes on the recent Arlington - Salem tests. Proc. IEEE **87**(11) (November 1999)

73. Cyclopedia of applied electricity. Technical report, Prepared by a Corps of Electrical Experts, Engineers, and Designers of the Highest Professional Standing, American Technical Society, Chicago, USA (1912)

74. P. Valdemar, System for producing continuous electric oscillations. *Transactions of the International Electrical Congress, St. Louis, MO, USA*, 2, September 12 (1904)

75. C.F. Elwell, The Poulsen system of wireless telephony and telegraphy. J. Electr. Power Gas 293–297, April 2 (1910)

76. J.S. Belrose, Fessenden and early history of radio science. *Proceedings of Radio Club of America* (November 1993)

77. A. Wehnelt, Ein Elektrolytischer Stromunterbrecher. *elektro-tech.Z.20*, pp. 76–79, January 26 (1899)

78. C.H. Heckler Jr., J.A. Baer, Research on reliable and radiation insensitive pulse-drive sources for all-magnetic logic systems. Project no. 3729, contract 950104 under nasw-6, Stanford Research Institute (SRI), Menlo Park, California (June 1962)

79. H. Armagnat, *The Theory, Design and Construction of Induction Coils (Translated and Edited By Otis Allen Kenyon)* (Mc-Graw Publishing Company, New York, USA, 1908)

80. L.D. Bliiss, *Theoretical and Practical Electrical Engineering* (The Bliss Electrical School, Takoma Park, Washington D.C., USA, 1922)

81. H.S. Norrie (Norman H Schneider), *Induction Ciols: How to Make, Use and Repair Them* (Spon & Chamberlain, 12 Cortlandt, New York, USA, 1901)

82. F.H. Beaudoin, *The Father of Wireless Telephony Reginald Aubrey: Fessenden Abridgement of the Biography of Fessenden (1866–1932)*. Cultural Committee of Austin, Printed by M. Leblanc Imprimerie, Austin, Texas, USA (February 2005). ISBN 2-923381-00-9

83. J.S. Belrose, Reginald Aubrey Fessenden and the birth of wireless telephony. Antenna's Propag. Mag. **44**(2) (April 2002)

84. R.A. Fressenden, How ether waves really move. Popular Radio **IV**, 337–347, November 5 (1923)

85. T.K. Sarkar, R. Mailloux, A.A. Oliner, M. Salazar-Palma, D.L. Sengupta, *History of Wireless*. (Wiley Interscience, John Wiley & Sons Inc. Publications, Hoboken, New Jersey, USA, 2006)

86. R.A Fressenden, *Wireless Telephony*. 25th Annual Convention of American Institution of Electrical Engineers, Atlantic City New Jeresy, USA (1908)

87. J.S. Belrose, A Fessenden's christmas eve broadcast retrospective. IEEE Antennas Propag. Mag. **54**(4) (August 2012)

88. A.M. Mayer, Lecture notes on physics (Part-I) mechanics, physics and chemistry. J. Frankl. Inst. **84**(5), 321–329 (November 1867)

89. A.M. Mayer, Researches in acoustics-Part 6. Phil. Mag. **49**, 352–365 (May 1875)

90. A.M. Mayer, On the minute measurements of modern science. Sci. Am. Suppl. **XXXVI**(8), February 24 (1877)

91. M. Leblanc, Etude sur le telephone multiplex. La Lumiere Electrique (Paris) **20**, April 17 (1886)

92. M. Hutin, M. Leblanc, Electric method of amplitude modulation, December 27 (1892). British Patent No. 23892

93. L. Rayleigh, *The Theory of Sound*, vol. 1, 2nd edn. (Macmillan, London, UK, 1894)

94. F.L. de Moura, Wave transmitter, October 11 (1904). https://patents.google.com/patent/US836531. US Patent No. 771917

95. F.S. Flosi, Father Landell de Moura radio broadcasting pioneer. Philatelia Chimica Et Physica **34**(1) (2012)

96. A. Hamilton, *The Other Side of Telecommunications - The Saga of Father Landell*, 1st edn. (Sulina Publishing, Porto Alegre, RS, Brazil, 1983)

97. F. Ernani, *The Incredible Father Landell de Moura*, 2nd edn. (Army Library Publishing, Rio de Janeiro, RJ, Brazil, 1984)

98. J. Maxwell, W6cf-amateur radio: 100 years od discovery. Technical report, QST-Copyright American Radio Relay League Inc. (January 2000)

99. T.H. White, Pioneering amateurs (1900–1917). United States Early Radio History, September 30 (1996)

100. J. Davy, (Brother of Sir Humphry Davy), *Memoirs of life of Sir Humphry Davy*, vol. 1 (Green and Longman, Paternoster-Row, London, UK, 1836). Printed by Sopttiswoode, New Street Square, London

101. J. Davy, The collected works of Sir Humphry Davy: Early miscellaneous papers from 1799 to 1805. With an introductory lecture and outlines of lectures on chemistry, delivered in 1802 and 1804

102. Sir J.M. Thomas, P.P. Edwards, V.L. Kuznetsov, *Sir Humphry Davy: Boundless Chemist, Physicist, Poet and Man of Action*, vol. 9 (Wiley-VCH Verlag GmbH & Co., KGaA, Weinheim, Germany, 2008)

103. Y. Yates, In the beginning: 10 inventors of the incandescent lightbulb. Technologist (2011)

104. T.H. White, A bulb of light? Sandy Mountain Historical & Technological Society, September 9 (2014)

105. A.L. Reimann, *Thermionic Emission* (Chapman and Hall Ltd., 11 Henrietta Street, WC 2, London, UK, 1934)

106. F. Guthrie, *Magnetism Electricity* (William Collins, Sons & Company, 1876). Specific Pages Referred: Platinum coil heated by current (284-288) and Cabon arc (299)

107. H.F. Dylla, S.T. Corneliussen, John Ambrose Fleming and the beginning of electronics. Technical report, Thomas Jefferson National Accelerator Facility (Jefferson Lab), 12000 Jefferson Avenue Newport News, VA 23606, USA (2004)

108. Vice Admiral H.G. Bowen, C.F. Kettering, *Edison Effect*. Technical report (The Thomas Alva Edison Foundation Inc., West Orange, New Jersey, USA, 2004)

109. J. Mitchell, I. Boyd, Prof Sir John Ambrose Fleming, F.R.S. Fleming valve centenary celebration 1904 - 2004:100 years of electronics. Technical report, Department of Electronic and Electrical Engineering, University College London, June 30 and July 01 (2004)

110. J.A. Fleming, Instrument for converting alternating electrical currents into continuous current, November 07 (1905). https://patents.google.com/patent/803684. US Patent No. 803684

111. L. De Forest, The Audion-I, A new receiver for wireless telegraphy. *Scientific American Supplement No. 1665*, pp. 348–350, November 30 (1907)

112. J. Elster, H. Geitel, Ueber die electricität der flamme. berichtigung. Annalen der Physik **252**(6), 193–222 (1882)

113. J. Elster, H. Geitel, Ueber electricitätserregung beim contact von gasen und glühenden körpern (on the generation of electricity by the contact of gases and incandescent bodies). Annalen der Physik und Chemie, 3rd series **97**(II), 1175–1189 (May 1883)

114. J. Elster, H. Geitel, Ueber die unipolare leitung erhitzter gase (on the unipolar conductivity of heated gases). Annalen der Physik und Chemie, 3rd series, pp. 315–329 (May 1885)

115. J. Elster, H. Geitel, Ueber die electrisirung der gase durch glühende körper (on the electrification of gases by incandescent bodies). Annalen der Physik und Chemie, 3rd series, pp. 109–127 (1887)

116. J. Elster, H. Geitel, Ueber die electricitätserregung beim contact verdünnter gase mit galvanisch glühenden drähten (on the generation of electricity by contact of rarefied gas with electrically heated wires). Annalen der Physik und Chemie, 3rd series, pp. 315–329 (1889)

117. J.J. Thomson, On the masses of the ions in gases at low pressures. Lond. Edinb. Dublin Philos. Mag. **48**(295), 547–567 (1899)

118. J.J. Thomson, *The Corpuscular Theory of Matter* (Charles Scribner's Sons, New York, USA, 1907)

119. O.W. Richardson, Monographs edited by J.J. Thomson. *Emission of Electricity from Hot Bodies* (Longmans Green and Co., 39 Paternoster Row, London, UK, May 01 1916)

120. J.J. Thomson, Carriers of negative electricity, Joseph J. Thomson's nobel lecture. Technical report, December 11 (1906)
121. G. Thomson, The electron. Technical Report Card Number: 68-62126, United States Atomic Energy Commission Office of Information, Services Library of Congress Catalog (1972)
122. J.J. Thomson, The modern theory of electrical conductivity of metals. J. Inst. Electr. Eng. **38**(183) (June 1907)
123. E. Riecke, Erst in verhältnismässig neuer zeit ist durch die elektronentheorien der metalle. Wied. Annal **66**, 560 (1898)
124. P. Drude, Zur elektronentheorie der metalle. Annalen Der Physik Vierte (Quarterly), Episode: Band 3 (1900)
125. L. De Forest, The Audion-II: a new receiver for wireless telegraphy. *Scientific American Supplement No. 1666*, pp. 354–356, December 7 (1907)
126. L. De Forest, *Father of Radio: The Autobiography of Lee de Forest* (Wilcox & Follett, 3 Westbrook Corporate Center, Suite 200, Westchester, IL 60154, 1950)
127. T.H. Lee, *The Design of CMOS Radio-Frequency Integrated Circuits (Chapter1: A Nonlinear History of Radio)* (Cambridge University Press, Cambridge, UK, 1998)
128. P. Vischer, How invisible lifeline rescues men from the sea. Popular Sci. Mon. 11–12 (January 1926)
129. C.H. Sterling (ed.) *Fritz Messere: Regulation, Encyclopedia of Radio*, vol. 1, 2, 3 (Imprint of Taylor and Francis Group, 29 West 35th street New York, 10001, USA, 2004)
130. M.C. Keith, C.H. Sterliing, C. O'Dell, *The Concise Encyclopedia of American Radio* (Routledge, Taylor and Francis Group, 270 Madison Avenue, New York, 10016, USA, 2010)
131. Regulations governing radio communication. Department of Commerce and Labor Bureau of Navigation, Government Printing Office, Washington, USA, September 28 (1912)
132. J. Schwoch, *The American Radio Industry and Its Latin American Activities 1900–1939* (University of Illinois Press, Urbana and Chicago, Library of Congress Cataloging, 1955)
133. F. Lowenstein, Telephone-Relay, April 24 (1912). https://patents.google.com/patent/US1231764. US Patent No. 1231764
134. W.D. Coolidge, Tungsten and method of making the same for use as filaments of incandescent electric lamps and for other purposes, December 13 (1913). https://patents.google.com/patent/1082933A. US Patent No. 1082933A
135. I. Langmuir, The effect of space charge and residual gases on thermionic currents in high vacuum. Phys. Rev. **2**, 450–486 (December 1913)
136. C.D. Child, Discharge from hot cao. Phys. Rev. **32**, 492–511 (December 1911)
137. I. Langmuir, The pure electron discharge. Proc. IRE **3**, 261–293 (September 1915)
138. E.H. Armstrong, Operating features of the Audion. Electr. World **64**, 1149–1151 (1914)
139. E.H. Armstrong, Some recent developments in the Audion receiver. Proc. Inst. Radio Eng. (IRE) **3**(4), 215–260 (September 1915)
140. E.H. Armstrong, Armstrong papers, from 1909 to 1956: the historical and interpretive collections. Collection, The Franklin Institute, September 12 (2016)
141. P.F. Godley, Armstrong's super-regenerative circuit, a discussion of its advantages, limitations and some of its variations, from the standpoint of assembly and operation, pp. 426–432 (September 1922)
142. C. Kitchen, High performance regenerative receiver design. *QEX*, Nov/Dec (1998)
143. M.C. Teich, Heterodyne systems and technology Part-1, in *NASA Conference Publication 2138*, New York, USA, March 12–27 1980. Columbia Radiation Laboratory, Columbia University. Work supported by the Joint Services Electronic Program of US Navy, US Army and US Air Force, under contract DAAG29-79-C-0079
144. J.L. Hogan, The heterodyne receiving system and notes on recent Arlington-Salem tests. Proc. IRE **1**, 75–97 (July 1913)
145. B. Liebowitz, The theory of heterodyne receivers. Proc. IRE **3**, 215–247 (September 1915)
146. J.W. Klooster, *Greenwood Icons of Invention: The Makers of Modern World from Gutenberg to Gates* (Greenwood Press, An Imprint of ABC-CLIO, LLC, 130, Cremona Drive, PO Box 1911, Santa Barbara, California, 93116-1911, USA, 2009)

147. E.H. Armstrong, A study of heterodyne amplification by the electron relay. Proc. Inst. Radio Eng. 145–168, April 5 (1917). Presented before I.R.E. on October 4 (1916)

148. E.H. Armstrong, The super-heterodyne-its origin, development, and some recent improvements. Proc. Inst. Radio Eng. **12**(5) (October 1924)

149. D.G. Godfrey, F.A. Leigh, *Historical Dictionary of American Radio* (Greenwood Press, Westport, Connecticut, USA, 1998)

150. E.H. Armstrong, A new system of short wave amplification. Proc. Inst. Radio Eng. 3–11 (February 1921)

151. C.H.J. Round, Direction and position finding. J. Inst. Electr. Eng. **59**(289), 224–257 (March 1920)

152. M.C.A. Latour, Marius latour audion or lamp relay or amplifying apparatus, February 7 (1922). https://patents.google.com/patent/US1405523. US Patent No. 1405523

153. R.E. Priestley, *The signal service in the European war of 1914 to 1918 (France)* (The Secretary, The Institution of Royal Engineers and The Signals Association, W. & J. Mackay & Co., Limited, Chatham, Canada, 1921)

154. P. Gannon, WW1: First world war communications and the tele-net of things. E&T: Eng. Technol. June 16 (2014)

155. M.A.C. Fuller, The fullerphone, and its practical application to military and civil telegraphy. IEE J. (1919)

156. M.A.C. Fuller, Fullerphone Mark-III, Pamphlet No. 3, Signal Training, Volume-III. Imperial House, Kingsway, London, WC2, UK, 1923. His Majesty's Stationary Office

157. L. Meulstee, Fullerphone principle of operation. Wireless for Warrior, May 8 (1919). http://www.wftw.nl/fullerpr.html

158. M.O.E. Buckley, *The Collection of E.H. Armstrong: Method of Reception Disclosed to Major O.E. Buckley* (Columbia University's Rare Book & Manuscript Library (RBML), May–June 1918)

159. H.W. Houck, The Armstrong super-autodyne amplifier. Radio Amateur News, The 100% Wireless Magazine (February 1920)

160. Armistice signed; end of war; Berlin seized by revolutionists; new chancellor begs for order; ousted Kaiser flees to Holland. The New York Times, November 11 (1918)

161. P.F. Godley, The far call: the phenomenal success of the amateur trans - Atlantic tests. Wirel. Age **9**(6) (March 1922)

162. J.B. Williams, *The Electronics Revolution: Inventing the Future* (Springer Praxis Publishing, Chichester, UK, 2017)

163. B. Jaker, F. Sulek, P. Kanze, *The Airwaves of New York: Illustrated Histories of 156 AM Stations in the Metropolitan Area 1921–1996* (McFarland & Company Inc. Publishers, Box 611, Jefferson, North Carolina, 28640, USA, 1998)

164. G. Davey, *Fun With Radio*. Fisrt Publisher: Edmund Ward (Publishers) Ltd. Revised Editions Publisher: Kaye & Ward Limited, 21 New Street, London EC2M 4NT, UK, 1957 (First), 1969 (Revised Editions)

165. D. Marc, Broadcasting, radio and television. Microsoft Encarta Online Encyclopedia, All rights reserved Microsoft Corporation, 1993–2000. http://encarta.msn.com

166. M.L. Sievers, *Crystal Clear Volume 1: Vintage American Crystal Sets, Crystal Detectors, and Crystals*, vol. 1 (The Vestal Press Ltd., 13851-0097, P.O. Box 97, Vestal, NY, United States of America USA, 1991)

167. M.L. Sievers, *Crystal Clear Volume 1: Vintage American Crystal Sets, Crystal Detectors, and Crystals*, vol. 2 (Sonoran Publishing, LLC, 85226 Chandler, AZ, 116 N Roosevelt, Ste. 121, United States of America (USA), 1995)

168. The construction and operation of a very simple radio receiving equipment. Technical Report Circular of the Bureau of Standards, No. 121, September 12, 1923, USA, Department of Commerce Bureau of Standards Washington, March 16 (1922)

169. A. Douglas, The legacies of Edwin Howard Armstrong. Proc. Radio Club Am. **64**(3) (November 1990)

170. G. Malanowski, *The Race for Wireless: How Radio was Invented (or Discovered)* (Author House^{TM}, 1663 Library Drive, Bloomington, IN 47403, USA, 2011)

171. J.F. Corrigan. *The P.W: Crystal Experimenter's Handbook*. Written for Popular Wireless (1925)

172. A. Douglas, *Radio Manufacturers of the 1920s*, vol. 2 (Sonoran Publishing, LLC, Chandler, Arizona-85226, USA, 1989). Freed Eisemann to Priess

173. A. Harry, O.P. Taylor, *Crystal Apparatus*, February 25 (1930). https://patents.google.com/patent/US1748435. US Patent No. 1748435

174. L.A. Hazeltine, Tuned radio frequency amplification with neutralization of capacity coupling. Proc. Radio Club Am. **2**(8), 7–12 (March 1923)

175. G.L. Archer, *History of radio to 1926*. Technical report (The American Historical Society, New York, 1938)

176. E.P. Wenaas, *Radiola: The Golden Age of RCA, 1919–1929* (Sonoran Publishing LLC, Chandler, Arizona, 85226, USA, 2007)

177. Radio service bulletin. Technical report, Department of Commerce Conference on Radio Telephony, May 1 (1922)

178. Radio service bulletin. Technical report, Department of Commerce Conference on Radio Telephony, April 2 (1923)

179. H. Hoover, *Recommendations for Regulation of Radio, Proceedings of the Third National Radio Conference* (Washington, D. C., USA, October 6–10, 1924)

180. H. Hoover, *Recommendations for Regulation of Radio, Proceedings of the Fourth National Radio Conference* (Washington, D. C., USA, October 9–11, 1925)

181. *RCA -Victor Service Notes Volume I : Radio Receivers Phonographs Television (1923–1937)*, vol. I, 1st edn. (RCA Victor Division of Radio Corporation of America, Camden, N.J., USA, 1944). This volume covers Notes previously issued for the years 1938 to 1942 inclusive

182. A. Douglas, *Radio Manufacturers of the 1920s*, volume 3: RCA to Zenith (Vol. 3 of 3), 1st edn. (The Vestal Press Ltd, 13851 Vestal, NY, P.O. Box 97, USA, 1914)

183. *Aircraft Year Book* (Aeronautical Chamber of Commerce of America, Inc., New York City, USA, 1927)

184. *Aircraft Year Book* (Aeronautical Chamber of Commerce of America, Inc. by D. Van Nostrand Company, Inc., 250 Fourth Avenue New York, USA, 1930)

185. G.P. Oslin, *The Story of Telecommunications* (Mercer University Press, 1501 Mercer University Drive Macon, Georgia, GA 31207, USA, reprint edition, 1999). ISBN 0865546592, 9780865546592

186. G.E. Sterling, R.B. Monroe, *The Radio Manual* (D. Van Nostrand Company, Inc., 120 Alexander St., Princeton, New Jersey (Principal office) 24 West 40th Street, New York 18, New York (USA), Fourth edition (April 1950)-Fifth printing (April 1960) edition, 1928)

187. H.T. Rogers, SE-1420, IP-501 & IP-501-A: the classic shipboard wireless receivers. Technical report, Western Historic Radio Museum, P.O. Box 511 109 South F Street Virginia City, NV 89440 (March 2018)

188. A.A. Huurdeman, *The Worldwide History of Telecommunications* (Wiley-Interscience, A John Wiley & Sons Inc. Publication, Hoboken, New Jersey, USA, 2003)

189. F.C. Bakewell, *Electric Science: It's History, Phenomena, and Applications* (Ingram Cooke and Co., London, UK, 1853)

190. Constantin M. SENLECQ, T électroscope. Les Mondes, revue hebdomadaire des sciences et de leurs applications aux arts et á l'industrie, par M. l'abbé Moigno **XLVIII**(13), 16 janvier (1879)

191. Constantin M. SENLECQ, T électroscope. Nature, Wkly. Illus. J. Sci. Paris **XIX**(482), January 23 (1879)

192. G.R. Carey, Transmitting, recording and seeing pictures by electricity. Electr. Eng. 57–58, January 16 (1895)

193. J.P. Gassiot, Bakerian lecture: on the stratifications and dark band in charges observed in torricellian vacua. Philos. Trans. R. Soc. Lond., R. Soc. Publ. UK **148**, 1–16 (1858)

194. G. Karsten, *Plücker, Julius, Allgemeine Deutsche Biographie (ADB) (in German)* (Duncker & Humblot, 26, Leipzig, Germany, 1888)
195. J.J. Thomson, *Conduction of Electricity Through Gases* (The University Press, Cambridge, UK, 1903)
196. R. Noakes, Cromwell Varley FRS, electrical discharge and Victorian spiritualism. Notes Rec. R. Soc. Lond. **61**, 5–21 (2006)
197. G. Thomson, Notes and records. R. Soc. Lond. **25**(2), 237–242 (December 1970)
198. W. Crookes, The Bakerian lecture-on the illumination of lines of molecular pressures and the trajectory of molecules. Philos. Trans. R. Soc. Publ. Lond. **170**, 135–164 (January 1879)
199. E. Goldstein, Vorläufige Mittheilungen Über elektrische Entladungen in verdünnten gasen. Technical report, Monatsberichte der Königlich Preussischen Akademie der Wissenschaften zu Berlin, May 04 (1876)
200. E. Goldstein, Über eine noch nicht untersuchte Strahlungsform an der Kathodeinducirter Entladungen. Technical Report Monatsber. II, Akademie der Wissenschaften zu Berlin (1886)
201. J. Mattingly, The replication of Hertz's cathode ray experiments. Study Hist. Philos. Modern Phys. Elsevier Science Ltd. **32**(1), 53–75 (2001)
202. J.J. Thomson, Cathode rays. The Electrician **39**(104) (1897)
203. L. Philipp, Über die electrostatischen eigenschaften der kathodenstrahlen. Annalen der Physik(AdP) **64** (1898)
204. W. Röntgen, Weitere beobachtungen über die eigenschaften der x-strahlen. Technical report, Mathematische und Naturwissenschaftliche Mitteilungen aus den Sitzungsberichten der Königlich Preussischen Akademie der Wissenschaften zu Berlin (1897)
205. F. Braun, Ueber ein verfahren zur demonstration und zum studium des zeitlichen verlaufs variabler ströme. Annalen der Physik und Chemie **60**(3rd series), 552–559 (1897)
206. A.A. Campbell-Swinton, The effects of strong magnetic field upon electric discharges in vacuo. Proc. R. Soc. Lond. **60**(359–367), 179–182 (1897)
207. A.A. Campbell-Swinton, Electric television. Nature, Nature Publishing Group **60**, June 18 (1908)
208. Who invented the television? how people reacted to john logie baird's creation 90 years ago. The Telegraph, January 26 (2016)
209. D.G. Godfrey, *C. Francis Jenkins: Pioneer of Film and Television* (University of Illinois Press, Urbana, USA, 2014)
210. J.R. Carson, Notes on the theory of modulation. Proc. I.R.E. **10**(1), 57–64 (February 1922)
211. E.H. Armstrong, A method of reducing disturbances in radio signaling by a system of frequency modulation. Proc. I.R.E. **24**(5) (May 1936)
212. J.R. Carson, T.C. Fry, Variable frequency electric circuit theory with application to the theory of frequency modulation. Bell Syst. Tech. J. **16**(4), 513–540 (October 1937)
213. C.H. Sterling, M.C. Keith, *Sounds of Change: A History of FM Broadcasting in America.* (University of North Carolina Press, September 15 2009)
214. D.A. Ferre, Early days of TV: what ever happened to channel 1? Radio Electron. Mag. **53**(3) (March 1982)
215. D. Jarvis, S.W. White, C. Wilson, M. Woody, *Images of America: Detroit Police Department* (Arcadia Publishing, Charleston SC, Chicago IL, Portsmouth NH, San Francisco CA, USA, 2008)
216. A short history of radio: with an inside focus on mobile radio. Technical report, Federal Communication Consortium (FCC), Winter (2003–2004)
217. J.A. Poli, Development and present trend of police radio communications. J. Crim. Law Criminol. **33**(2) (1943)
218. The first police radio communication system. Random Historical Snippets of The World of Nerds, Spark, Aether, and other Wireless 'Thingamabobs', June 23 (2017)
219. A timeline overview of Motorola history: 1928–2009. Motorola, Jan 05 (2015)
220. Research Assistant Guy L. Cramer and grandson of Donald Lewes Hings. Donald lewes hings: Development of walkie talkie from 1930–1945. presented to the National Research Council of Canada, August 31 (2001)

221. W.E. Crook, *D/F Hand Book for Radio Operators* (Sir Isaac Piman and Sons, Ltd., Pitman House, Parker Street, Kingsway, London WC2, UK, 1942)
222. D.J. Cichon, W. Wiesbeck, The Heinrich Hertz wireless experiments at Karlsruhe in the view of modern communication, in *100 Years of radio*, number Conference Publication, 0537-9989; No. 411, International Union of Radio Science, Institution of Electrical Engineers IEE. Science, Education and Technology Division, British Vintage Wireless Society, Savoy Place, London, England, September 5–7 (1995). Institution of Electrical Engineers Conference
223. A.S. Popov, Apparatus for the detection and recording of electrical oscillations (in Russian). Zh. Russ. Fiz.-Khim. Obshchestva (Physics, Pt. 1) **28**, 1–14 (1896)
224. A.S. Popov, To the editor. The Electrician **40**, 235 (December 1896)
225. J.S. Stone, Method of determining the direction of space telegraph signals, December 16 (1902). https://patents.google.com/patent/US716134. US Patent 716134
226. W. Holpp, The century of radar-from christian hülsmeyer to shuttle radar topography mission Technical report, Defense Electronics, EADS Deutschland GmbH, Ulm/Donau Germany (2004)
227. The telemobiloscope. Electr. Mag. **2**, 388 (1904)
228. J. Ender, 98 years of the radar principle: the inventor christian hülsmeyer (exhibits at the deutsche museum munich), hülsmeyer memorial speech in the town hall of cologne, in *EUSTAR-2002*, June 04 (2002), http://www.design-technology.info/resourcedocuments/Huelsmeyer_EUSAR2002_english.pdf
229. F. Nebeker, *Dawn of the Electric Age: Electrical Technologies in the Shaping of the Modern World, 1914 to 1945* (A John Wiley & Sons Inc., Publication, Hoboken, New Jersey, USA, 2009). ISBN 978-0-470-26065-4
230. E. Bellini, System of directed wireless telegraphy, December 21 (1909). https://patents.google.com/patent/US943960. US Patent 943960
231. H. Norinder, *Ch4: Long-Distance Location of Thunderstorms in Thunderstorm Electricity by H. R. Byers* (University of Chicago Press, Chicago, USA, 1953)
232. Weather radar. Report Fact Sheet No 15, Met Office, National Meteorological Library and Archive, UK, July 2007. Crown Copyright
233. F. Adcock, Improvements in means for determining the direction of a distant source of electro-magnetic radiation. European Patent Office and Google, August 20 (1918). https://www.epo.org. GB Patent 130490, [Online; accessed 27-May-2018]
234. N.J. Willis, H.D. Griffiths, *Advances in Bistatic Radar*. Scitech, SciTech Publising Inc., Raleigh, 911, Paverstone Dr.-Ste B, NC-27613(USA), ISBN 1891121480. ISBN13: 9781891121487 (2007)
235. Y. Blanchard, A French Pre-WW-II Attempt at Air-Warning Radar: Pierre David's "Electro-magnetic Barrier". *The Radio Science Bulletin* **358**, 18–34 (September 2016)
236. ITD History. https://www.nrl.navy.mil/itd/Overview. [Online; accessed 18-July-2018 at 10:23 IST]
237. W.F. Blanchard, Air navigation systems. J. Navig. **44**(3), 285–315 (September 1991)
238. E.V. Appleton, M.A.F. Barnett, On some direct evidence for downward atmospheric reflection of electric rays. Proc. R. Soc. **109**, 261–641 (December 1925)
239. M. Rothenberg, *The History of Science in the United States: An Encyclopedia* (Garland Publishing Inc., 29 West, 35th street, New York, 10001, 2001). ISBN 0-203-90280-7. An imprint of Taylor and Francis Group
240. P. Ptak, *Aircraft Tracking and Classification with VHF Passive Bistatic Radar*. Ph.D thesis, Department of Mathematics and Physics, Lappeenranta University of Technology, Finland (June 2015)
241. P.C. Carter, F.E. Nancarrow, A.C. Mumford, H.T. Mitchell, Interference by aeroplanes. Gp radio report, 233, part v, British Post Office, June 03 (1932)
242. J. Darricau, *Physique et Théorie du Radar*, vol. 1, 2 and 3, 2nd edn., Paris (February 1981). ISBN 978-2-9544675-1-1. Chapter-1: Histoire du Radar dans le Monde puis France by Jacques DARRICAU et Yves BLANCHAR

243. G. Massimo, The early history of radar. IEEE Industr. Electron. Mag. **4**(3), 36–42 (2010). October

244. F.A. Kolster, F.W. Dunmore, The radio direction finder and its application to navigation, in *Scientific Papers of the Bureau of Standards*, vol. 17, pp. 529–564 (National Institute of Standards and Technology, 100 Bureau Drive Mail Stop: 2000 Gaithersburg, MD 20899, USA, July 09 1921)

245. R. Keen, *Direction and Position Finding by Wireless* (The Wireless Press Ltd., 12 & 13, Henrietta Street, Strand, London WC2, UK, 1922)

246. Captain H.J. Round, electronics-notes.com. https://www.electronics-notes.com/articles/ history/pioneers/captain-h-j-round.php. [Online; accessed 04-June-2018]

247. G.B. Mason, World War 2 at Sea: HF/DF or HUFF DUFF - High frequency radio direction finding in royal navy warships. http://www.naval-history.net/xGM-Tech-HFDF.htm (1992). [Online; accessed 28-May-2018]

248. G. Helgason, *Weapons and Technologies - Fighting the U-boats: hf/df The High Frequency Direction Finder*. U-boot Archive: Deutsches U-Boot Museum, Raudalaekur 34, IS-105 Reykjavik Iceland. https://uboat.net/allies/technical/hfdf.htm. [Online; accessed 27-May-2018]

249. J. Clarke, Homing radio: a directional device for guiding pilots to airports. Wonders of World Aviation **1,2,3**(Part-9), March 22 (1938). http://www.wondersofworldaviation.com/homing_ radio.html. [Online; accessed 27-May-2018]

250. O. Blumtritt, H. Petzold, W. Aspray (eds.), *Tracking the Histrory of Radar*. A Publication of IEEE-Rutgers Centre for History of Electrical Engineering and the Deuteches Musium, Institute of Electricals and Electrronics Engneers, 445 Hoes lane, P.O. Box 1331, Piscataway, New Jersey 08855-1331 (908)932-1066 (1994)

251. B.A. Austin, Precursors to radar the Watson-Watt memorandum and the Daventry experiment. Int. J. Elect. Eng. Educ. **36**, 365–372 (1999)

252. D.L. Boslaugh, *When Computers Went to Sea: The digitization of United States Navy* (The Institute of Electrical and Electronics Engineers, IEEE Service Centre, 445 Hoes Lane, PO Box No. 1331, Piscataway, NJ 08855-1331, USA, 1999)

253. B. Austin, Daventry experiment: the birth of British radar. Radio Bygones (153), 28–35 (February/March 2015)

254. T.E. Derham. *Design and Evaluation of a Coherent Multistatic Radar System (UMI U593583)*. Ph.D thesis (Department of Electronic & Electrical Engineering, University College London, 789 East Eisenhower Parkway, P.O. Box 1346, Ann Arbor, Ml 48106-13, USA, April 2005)

255. N.J. Willis, Bistatic radars and their third resurgence: passive coherent location, in *IEEE Radar Conference*, Long Beach, USA (April 2002)

256. N.J. Willis, *Bistatic Radar*. Scitech (SciTech Publishing Inc., Raleigh, 911, Paverstone Dr.-Ste B, NC-27613 (USA), 2005)

257. H.D. Griffiths, C.J. Baker, *An Introduction to Passive Radar* (Artech House, 16 Sussex Street, London, SW1V 4RW, (UK), 2017)

258. E. Ali, A. Örstadius, Passive radar detection of aerial targets. Technical report, Lund University, Lund, Sweden, May 23 (2017)

259. S. Stromoy, *Hitchhiking Bistatic Radar* (University of Oslo, Department of Physics, University of Oslo, Norway, May 21 2013)

260. V.S. Chernyak, *Fundamentals of Multisite Radar Systems: Multistatic and Multiradar Systems* (Gordon and Breach Science Publishers, 325 Chestnut Street, Suite 800, Philadelphia, PA 19106 United States, 1998). ISBN 90-5699-165-5

261. C. Pell, E. Hanle, Bistatic and multistatic radar. IEE Proc. **133**(7), 585–586 (December 1986)

262. P. Bezoušek, V. Schejbal, Bistatic and multistatic radar systems. Radioengineering **17**(3), 53–59 (September 2008)

263. M. Cherniakov, M. Gashinova, P. Lombardo, D. Pastina, M. Martorella, C. Baker, Bistatic and multistatic radar, in *The 13th Europian Radar Conference (EURAD-2016)*, London, UK, October 3–7 (2016)

264. Mike@bitaboutbritain. What really happened at orford ness? http://bitaboutbritain.com/what-really-happened-at-orford-ness/

265. A. Chodos, April, 1935: British patent for radar system for air defense granted to Robert Watson-Watt. APS News: Month in Physics History, APS Phys. **15**(4), 2 (April 2006)
266. S. Horncastle, Chain home, RAF canewdon: a history of one of the first radar stations. *Rochford District Community Archive*, November 18 (2013). www.rdca.org.uk
267. B.T. Neale, CH - The first operational radar. GEC J. Res. **3**(2), 73 (1985)
268. R. Rzemien, Coherent radar: editor's introduction. Ohns Hopkins APL Tech. Digest **18**(3) (1997)
269. G. Galati, *100 Years of Radaar* (Springer International Publishing AG, Switzerland, Switzerland, 2016). ISBN 3-319-00583-3
270. P. Whitney, Those funny old radio tubes. Antique Radio Classif. **10**(5), 17 (May 1993)
271. J.L. Baird, Aircraft to ground television equipment. Baird Television Equipment, July 21 (1939). Equipment supplied under contract to French Air Ministry
272. H. Griffiths, Some reflections on the history of radar from its invention up to the second world war. *Newsletter of the Jmes Clerk Maxwell Foundation* (8), Spring (2017). ISSN ISSN 2058-7503 (Print) and SSN 2058-7511 (Online)

Index

© The Author(s), under exclusive license to Springer Nature Singapore Pte Ltd. 2021   205
V. Patil, *Chronological Developments of Wireless Radio Systems before World War II*,
https://doi.org/10.1007/978-981-33-4905-6